AF740797

The Jewish Tradition, Sexuality and Procreation

Lewis D. Solomon

University Press of America,® Inc.
Lanham · New York · Oxford

Copyright © 2002 by
University Press of America,® Inc.
4720 Boston Way
Lanham, Maryland 20706
UPA Acquisitions Department (301) 459-3366

12 Hid's Copse Rd.
Cumnor Hill, Oxford OX2 9JJ

All rights reserved
Printed in the United States of America
British Library Cataloging in Publication Information Available

ISBN 978-0-7618-2227-1

™ The paper used in this publication meets the minimum requirements of American National Standard for Information Sciences—Permanence of Paper for Printed Library Materials, ANSI Z39.48—1984

Dedication

For my wife, Janet, and our son, Michael

Contents

Acknowledgments

I want to single out two individuals for their help. Beatrice Bruteau, Ph.D. read a complete draft of this book and gave me extensive comments that guided me in my revisions. Nell B. Taylor typed the manuscript and helped index the book.

Todd G.E. Melnick and John P. Miller, Reference Librarians at the Jacob Burn Law Library of The George Washington University Law School, assisted in locating hard-to-find materials. Rabbi Jan Caryl Kaufman and Rabbi David J. Fine of the Rabbinical Assembly provided useful research materials. I also benefitted from using the I. Edward Kiev Collection at The George Washington University's Melvin Gelman Library.

Chapter 1

Introduction

There is a spiritual emptiness at the core of the life of many Americans. In facing questions about sexuality and procreation, our condition is one of anxiety, of being lost, of searching for a spiritual home. Once we sought a "home" together through family and religion. We looked to religious authority for guidance, if not answers, to our questions and our children's questions about sexual and reproductive ethics.

Now, we are more likely to search alone, especially through sexual exploration. Yet, for many sex has become an anxious, arduous pilgrimage. Behind us is a path often littered with emotional wreckage marked by one heartbreak after another. Our search, furthermore, takes place in the context of a significant societal transformation.

Our society is marked by far greater diversity than ever before. There are new forms of individual and social life. We have opened American society to women and people of color. There is increased tolerance for people who differ from the norm, whether in race, religion, or sexual orientation. America is unlikely to return to the mid-twentieth century, a time of premarital chastity, low divorce, stay-at-home moms, or the closeting of homosexuals.

In the midst of these societal changes, most of us value our autonomy. We wish to let responsible individuals make their own choices, to chart their own sexual destinies and choose their own reproductive path, provided they do not violate others' rights or threaten society.

We shy away from judging others. We do not condemn two adults who consent to a sexual act. Most of us view sexual conduct and procreation as entirely one's own business.

With the loosening of behavioral boundaries, we have become far more ambivalent. Although many of us hunger for moral values and for clear guidelines, we resist imposing our standards, however tenuously, on others. In our discomfort with passing judgment on others, most of us strictly observe the Eleventh Commandment: Thou shall not judge; how others adhere to the first ten is a private matter.

We seek tolerance and civility; respect for our diversity of views and lifestyles is foremost in our minds. A tolerant society is deeply entrenched. We do not want to infringe on what others do. We are more willing to tolerate lifestyle experimentation, provided it is done safely and "moderately."

Yet, our choices about sexuality and reproductive matters often represent difficult, if not heart-wrenching, decisions. We face the serious challenge of how we think about ourselves and our children, about religion, ethics, and values. As we begin this inquiry into basic human relationships, each of us confronts the meaning and value of human life.

Although contraceptives and the availability of abortion often deprive sexual intercourse of any spiritual meaning, many of us perceive ourselves as moral actors with responsibilities, obligations, and concerns. At the same time, we are not willing to surrender our autonomy to absolute authority, obey received religious dictates, or to submit to an institutionalized faith. We no longer can find morality from the towering heights of Divine revelation or its embodiment in rigid moral codes. Many of us do not want to cosset ourselves by religiously mandated rules or norms.

We do not want a religious authority telling us how to lead our lives. Also, we want to let responsible individuals make their own choices. Yet, there is a spiritual hunger for some type of guideposts or boundaries, if not standards, in dealing with sexuality and procreation, for what is acceptable and unacceptable behavior. Many of us realize that appeals couched in terms of autonomy, often a code word for amorality with immediate sexual gratification the highest goal, no longer suffice. We are interested in looking to the Jewish tradition to set standards of "right" conduct. We desire our children to have proper values, to learn about a moral order based on spiritual coherence and structure.

In dealing with questions about sexuality and procreation, it is helpful to face these dilemmas informed, but not straightjacketed, by the Jewish tradition, one that manifests a long-standing aversion to patterns of human behavior, of loving and reproducing, that the ancients regarded as "unnatural." Judaism provides an ideal of how children are to be brought into the world, namely, through a male-female sexual union in a conjugal relationship. It encourages family stability. It tells us not to "shack up" or to have abortions "on demand;" not to do other things celebrated in the popular culture at every turn.

In examining sexuality and procreation, this book is unique in two respects. First, it presents a wide spectrum of viewpoints from various strands of contemporary Judaism -- traditional (Orthodox) and more modern and flexible (Conservative, Reform, and Reconstructionist). In providing the guidance offered by the Jewish tradition, I draw upon modern Jewish commentators, who, while looking to ancient sources, have witnessed the contemporary societal transformation over the past three or four decades.

Second, while some would follow the dictates of the traditional or more modern Jewish commentators and others would abandon any Jewish mooring, there is a need to rethink the Jewish tradition and its reliance on the commanding authority of biblical and rabbinic sources. In place of legalistic guidelines taking the form of a prescribed set of rules, this book offers the nondogmatic perspective of Spiritual Judaism, resting our faith in God and our ongoing relationship with the Eternal in improving the world and on our autonomy and the ability to choose an appropriate lifestyle, grounded in certain interpersonal virtues, namely, love, compassion, and personal responsibility. Spiritual Judaism presents a coherent moral vision enabling us to build on a foundation of autonomy, yet place our decisions about sexuality and procreation within a larger spiritual framework. This book strives to create an ethical system based on interpersonal virtues, derived from Jewish sources, ancient and modern.

Before offering an overview of this book, by way of background it is helpful to delve a little more deeply, but briefly, into America's moral and cultural transformation.

America's Moral and Cultural Transformation

From 1965 to 1995, the United States faced moral and social decline. Indicators of social dysfunction skyrocketed, including illegitimacy, childbearing by children, rampant divorce, crime, and unsafe streets. Something happened to American society over these thirty or so years.

A transformation in values occurred from, say, the mid 1960s to the late

1970s. Individualism in personal lifestyles soared; expressiveness and pleasure-seeking came to the fore accompanied by a demand for quick excitement. A cultural rebellion led to a defiance of traditional norms, marked by the devaluation of personal responsibility and traditional sexual morality. Many of us came to live for the moment.

The place of females in society changed. Women were freed from their traditional roles of stay-at-home moms by the advent of safe and effective contraceptives, household labor-saving devices, the reduction in infant mortality to nearly zero, and the rise of the service economy and the accompanying changes in the nature of work with mental labor replacing physical exertion. These brought about a profound shift in sexual behavior and the family structure.

The widespread use of birth control pills, perhaps the best representative of the sexual revolution, initiated a radical change in attitudes. Oral contraceptives made possible sex with nearly no fear of pregnancy. They sanitized sex from any reminders that procreation was an issue.

Women became free to decide their own reproductive future. Women could prevent or opt out of pregnancies.

At the beginning of the twenty-first century, we are witnessing a rather complicated rise and fall of moral standards.[1] It is a mixed bag; moral standards have surged in one area and fallen in another. Our notions about public drinking are likely more restrictive than at anytime since Prohibition. Fifty years ago, people drank at lunch with abandon, not so anymore. Our culture now stigmatizes, even nearly penalizes, behavior once taken for granted, such as, cigarette smoking, public celebration of drug use, anti-Semitism, and other forms of racial and ethnic disrespect.

At the same time, there has been coarsening, if not a degradation, in culture. Much of popular entertainment, in my view, ranges from the vulgar to the obscene. We live in a highly sexualized society, one that promotes insatiable sexual longings. There is a continuous flaunting of sensuality in comedy and on television. Unwed mothers serve as television and Hollywood heroines. The film industry, the music business, and the television networks willingly put their talents to work to give public standing to and help fuel the values of vulgarity, violence, and hostility to others.

Popular culture radiates blatant sexual appeals and encourages rampant sexual activity. Promiscuity on television sitcoms has a level of explicitness unthinkable thirty or forty years ago. Many are offended by the easy availability of pornography in print and on the internet, erotic advertising, the proliferation of four-letter words in movies, internet ads offering phone sex, the sadism of rap music, the exhibitionism and narcissism on talk shows, and the general decline in our reticence about sexual matters.

We are more tolerant of frank sexual talk and jokes in public. But we evidence an intolerance of ethnic jokes and are repulsed by any talk or conduct that could be viewed as sexual harassment.

Elsewhere, trends point to more conservative contemporary attitudes about sexuality and promiscuity. We look down on and want to eliminate the unhealthy, the unsafe. We use health codes to regulate our carnal desires. We want safe, responsible sex, based on a model of disease prevention.

In the 1970s and 1980s, people became more tolerant of divorce. As women became more independent, they demanded more of marriage; they became less dependent on husbands. They would not as easily put up with "unrewarding" marriages or "flawed" husbands. Men abandoned their families, leaving aging wives for ever-younger trophy wives, so-called "trade-ups." All these factors have reduced the stability of marriage and increased the prevalence of divorce. The family became brittle; families disintegrated. At the same time, the rate of unmarried cohabitation soared by over 50% in the decade of the nineties, after almost doubling from 1980 to 1990.[2]

Some things have improved over the past decade. Certain social indicators are moving in the right direction: divorce rates are dropping and teen pregnancies are down. Although the incidence of divorces remains high with about 40% of new marriages ending in divorce, the divorce rate (including annulments) declined in the United States, drifting downward from 4.7 per thousand people in 1990 to 4.3 in 1998 reflecting, in part, a trend toward marriage at older, and presumably more mature, ages.[3] Today, people view divorce as a graver tragedy than twenty-five years ago.

The teenage birthrate in the United States decreased over 20% between 1991 and 1999, from 62.1 to 49.6 births for every 1,000 girls, aged 15 to 19, as a result of less teen sex and more use of birth control devices. However, it still remains the highest in the developed world.[4]

Other social indicators are moving in a negative direction. There has been a notable surge in nonmarital childbearing in the 1990s. For millennia, it was "illegitimate" to have children outside of marriage. This age-old assumption has been overthrown, resulting in an astonishing rise in out-of-wedlock births.

In 1999, one third of all births in the United States were to unmarried mothers, up from 26% in 1990. Illegitimacy rates among white women (including Hispanics) in 1999 soared to 26.8%, a significant increase from the 16.9% recorded in 1990 (never mind the 3% in 1970),[5] reaching what many experts believe is a "tipping point."

The rise of single motherhood has taken place in the context of a sea change in attitudes toward sexual and family issues, including the status of females in society, the value or necessity of women working outside the house, premarital sex, the ideal of sexual freedom, the unreliability and irresponsibility of men, as well as accidents of fate and choice.[6] With the loosening of the sexual mores in the 1960s and 1970s, more women became saddled with fatherless children or were abandoned by men chasing after the next young, nubile sex object. Men were freed from the responsibility of dealing with the consequences of sex, the women they impregnated and the children they fathered. There are far fewer "shotgun" marriages today.

An Overview of the Book

To set the stage of our discussion of sexuality and procreation, Chapter 2 explores two Jewish approaches to these questions: one, a text-oriented methodology, and a second, focusing on our autonomy, grounded in our Godlike nature and our partnership with the Eternal in perfecting the world. In dealing with issues of sex, family structure, reproduction, and abortion, the text-oriented tradition often seeks to restrict our desire for personal autonomy.

In this chapter, I envision a transformation of Judaism based on virtues, not legalistic rules, as guideposts for human behavior. I develop the need to exercise our autonomy and pursue self-fulfilment, letting people be free to shape their approach to sexuality, procreation, and family life in a way that is meaningful to them, subject to certain criteria or norms, not binding rules. Following the principles of Spiritual Judaism, certain virtues, namely love and compassion, should guide our decision-making in matters of sexuality and procreation. Spiritual Judaism also emphasizes the importance of personal responsibility in our actions.

Chapters 3 and 4 strive to formulate a Jewish sex ethic. Unlike some other religions, Judaism declares sexuality a valued aspect of our humanity. However, the Jewish tradition sought to steer human sexuality, seen as a powerful force, away from both premarital sex and homosexuality and to channel this desire in a positive direction toward procreation by a married, heterosexual couple. As the late Lord Jakobovits, the former chief rabbi of the United Kingdom, put it, "To fulfill [one's] destiny as a creative creature in the image of his Creator, [a person] must be generated and reared out of the intimate love joining husband and wife together, out of identifiable parents who care for the development of their offspring, and out of a home which provides affectionate warmth and compassion."[7]

We cannot be blind, however, to contemporary social realities. Unmarried people engage in sexual activities, from teenagers with raging hormones, to adults deferring marriage in order to complete their education and get started in their careers, to seniors, widows and widowers, seeking companionship but unwilling to upset pension arrangements. Uncommitted sex, that is, having sex as an activity aimed at self-fulfillment, as well as the widespread practice of individuals living together in unmarried heterosexual relationships and the growing number of committed, ongoing homosexual couples, necessitate a reevaluation of the Jewish tradition's view regarding who may engage in sexual intercourse and when.

Sex, in all its specialness, makes a unique contribution to human existence. Spiritual Judaism approves of premarital sex, particularly for engaged couples, if characterized by the connection of sexuality and spirituality, enabling individuals to experience God in one's most intimate relation with another human being. Allowing individuals to decide at what point in an ongoing, loving relationship they are able to link sexuality and spirituality will facilitate self-examination and encourage personal growth and ever higher, more noble strivings. Let us learn how to experience spirituality, a sense of our wholeness, the interconnection of body-mind-spirit, through our sexuality, thereby realizing the Divine Presence in our lives and our spiritual essence.

Chapter 4 considers homosexuality in the Jewish tradition. Departing from the unanimous condemnation of the text-oriented approach, Spiritual Judaism, along with the Reform and Reconstructionist Movements, looks with approval on the ongoing, committed relationships of homosexual couples. For Spiritual Judaism, there is no "natural" process of sexuality or procreation.

Although Spiritual Judaism is committed to the flourishing of nuclear families, consisting of a husband and a wife and their children produced by post-marital intercourse, as an essential foundation for a well-functioning society, we are in the process of redefining our concept of the family to include not only heterosexual couples, but also gay and lesbian families, having children conceived in the context of same sex unions, and single parent households. These new family configurations produce offspring through the wonders of modern science, notably assisted reproduction techniques, including artificial insemination, various types of surrogacy arrangements, and in the future, human cloning, considered in Chapters 6 and 7. As developed in Chapter 4, the diversity of arrangements supports fundamental family values.

Chapter 5, in focusing on the vexing issue: "Who is a Jew?", presents the Jewish tradition's longstanding concern for generational integrity and genealogical lineage. During the last quarter of the twentieth century, a massive divide arose between Orthodox and Conservative Jews, on one hand, who continue to look to matrilineal descent, and, on the other hand, the Reform and Reconstructionist Movements accepting patrilineal descent, with certain conditions, in the context of interfaith marriages. By offering a nonbiological classification approach, Spiritual Judaism moves the dialogue forward. If someone indicates, "I am Jewish," this chapter concludes that we accept one's word for his or her religious status, subject to certain minimal conditions set forth in Chapter 5, without lineage, educational or upbringing inquiries, or an involved conversion process. Let those who declare themselves Jews be Jewish. The new ways in which conception occurs will further propel a reconfiguration of one's religious identity.

In discussing the diverse ways of procreation through modern technologies, Chapters 6, 7, and 8 force us to focus our attention on our understanding of life, specifically what it means to be human. Technological advances, such as manipulating sperm, poking eggs, or taking single cells out of embryos and testing them for genetic defects, redefine our sacrosanct notions of motherhood and family. They are redrawing our images of fertility and aging, transforming sex and procreation currently by permitting the creation of babies without sexual intercourse, through artificial insemination or in vitro fertilization.

Chapters 7 and 8 consider modern technological frontiers, cloning, separating new human existence from sexuality, and genetic testing and engineering, offering the possibility for controlling the quality of offspring and other characteristics, such as gender. Through cloning and genetic engineering we can look forward to creating individuals to specification. Each of us may clone to create a nearly perfect replica of ourselves. Genetic engineering will enable the creation of what we regard as a more "perfect" child. We can foresee a degree of perfectability, increased control over our offspring, their physical attributes and brains.

Parents' quest for beautiful genes for their offspring today involves payment to sperm or egg donors and genetic testing. The future promises cloning and genetic engineering. This trend represents an outgrowth of the urge that people have to mate with whom they perceive as genetically "superior" people and produce children with evolutionary advantages, such as height, beauty for women (or good looks for men), intelligence, and social skills.

Yet, we are only beginning to grapple, as Chapters 7 and 8 do, with what it means to create humans to specification. We also need to consider what it means for us to entertain the possibility of postponing death and nearly gaining earthly immortality.

Contemporary Jewish authorities, in responding to these advances on the reproductive frontier, have kept in mind that the Jewish tradition teaches us not to tread in God's domain. For example, the builders of the Tower of Babel after coming too close to heaven, spoke different languages (Genesis 11:1-9).

Decades ago, modern medical advances, such as test tube babies, faced initial criticism that humans were playing God, by tampering with the Divine arrangement for creation. With the growing acceptance of assisted productive techniques over the past two decades, human participation in the Eternal's domain is steadily increasing.

The use of future technological advances, such as cloning and genetic engineering, to create more perfect humans and design offspring to our specifications, with fewer physical and emotional problems, will face similar objections. These technological achievements raise the specter of humans exercising power over others and suggest that we are going where we do not belong; we are trying to be God and control the uncontrollable.

What will it mean for Judaism, when characteristics innate throughout all of human history become a matter of choice or manipulation? Running throughout Chapters 6, 7, and 8 is the theme that as co-creators with God in perfecting the world we are not barred from improving or changing "nature."

Today, single individuals and couples, heterosexual and homosexual, use (and will continue to use) modern technologies separating sex and reproduction to achieve their procreational goals. In the future, we should not limit or discourage the growing array of choices redefining and restructuring families. Freedom to use assisted reproductive technologies and, in the future, to clone humans supports the ability of infertile heterosexual couples, single individuals, and same sex couples to have and rear children.

Spiritual Judaism favors experimentation, based on the capacity of humans to use God-given knowledge, to grow and become better. It advises going ahead with the relentless march of technology. Do not doom humans to a state of misery. When human cloning and genetic engineering come to fruition, we still will remain beings in the image of God. Spiritual Judaism recommends letting people use this God-given knowledge, recognizing it is bestowed on us by the Eternal, for the betterment of humanity. Let it be used, however, subject to the virtues advocated by Spiritual Judaism, namely, love, compassion, and personal responsibility.

Chapters 7 and 8 advocate allowing individuals to determine what is best for them and their offspring. Spiritual Judaism suggests permitting individuals to make choices to help their children, so-called "designer" children, lead better lives. Through genetic engineering, parents will be able to prevent diseases and enhance their children. If we accept the rights of parents to control and benefit their offsprings' lives and identities through social and environmental influences they today regard as positive, likewise we ought to allow a wide array of genetic interventions, even genetic enhancement.

Chapter 9 confronts the troublesome and controversial topics of abortion and selective pregnancy reduction, placing them in a larger spiritual framework. Following the introduction of the birth control pills in the early 1960s and the rapid and unprecedented acceptance of abortion beginning in 1973, women in the United States were able to manage their reproductive futures. They could decide whether or not they wanted to get pregnant as well as whether or not to terminate the existence of a fetus.

Women's sexual autonomy, their freedom of choice, has assumed an important value. When women get pregnant, their lives change forever. They sought the right to control their own lives leading to "on demand" abortion.

There is an unhappy choice between reducing the desire for abortions and trying to protect the freedom of pregnant females. Concern about the status and autonomy of women impacts on the question of abortion. A need exists for women to have some level of control over their biological life if they are to assume a position of equality in society. However, the fetus is a potential person; thus, Spiritual Judaism proposes that such protection and control be exerted prior to conception. Spiritual Judaism, furthermore, does not support the notion of unfettered abortion simply "on demand" or at late points in a pregnancy.

The question of when a human embryo claims a right to life and thus should not be aborted remains difficult, if not insoluble. Spiritual Judaism looks to modern science to help answer this question. The best contemporary knowledge places the beginning of the higher brain function, connoting ensoulment, at twenty-four weeks, at the same time a fetus achieves viability. Although advocating responsibility in sexual intercourse and rejecting abortion "on demand" as a form of post-conception birth control, Spiritual Judaism would permit abortions for maternal or fetal health during the first and second trimesters of pregnancy; but thereafter only to save maternal life.

Abortion, together with female contraceptives, enable women to exit from nearly total male domination of their reproductive lives. However, with this freedom comes responsibility and a need for sex education. Prior to intercourse, couples ought to use effective birth control devices to prevent unwanted pregnancies, thus substantially decreasing the need for abortions.

Chapter 10 offers a brief conclusion. It supports the transformation of women's role in society and the family and favors the evolution of diverse family structures and the use of modern procreative technologies.

Notes

1. I have drawn on David Brooks' wonderful book, *Bobos in Paradise: The New Upper Class and How They Got There* (New York: Simon & Schuster, 2000), 190-203, 216-217.
2. U.S. Census Bureau, *Statistical Abstract of the United States: 2000*, 120th edition (Washington, DC: U.S. Government Printing Office, 2000), Table No. 57. Unmarried Couples by Selected Characteristics: 1980 to 1999, 52.
3. *Id.*, Table No. 150. Marriages and Divorces - Number and Rate by State: 1990 to 1998, 104.
4. Stephanie J. Ventura et al, "Births: Final Data for 1999," *National Vital*

Statistics Report 49:1 (Hyattsville, MD: National Center for Health Statistics, 2001), 28. Table 4. Total fertility rates and birth rates by age of mother: United States, 1970-99, and by age and race of mother: United States, 1980-99; Sheryl Gay Stolberg, "U.S. Birth Rate at New Low as Teen-Age Pregnancy Falls," *New York Times*, April 29, 1999, A22; Marc Lacey, "Teen-Age Birth Rate in the U.S. Falls Again," *New York Times*, October 27, 1999, A14.

5. Ventura et al, "Births: Final Data for 1999," 44, Table 17. Number, birthrate, and percent of births to unmarried women by age, race, and Hispanic origin of mother: United States, 1999; 45-46, Table 18. Birth rates for unmarried women by age of mother: United States, 1970, 1975, and 1980-1999, and by age, race, and Hispanic origin of mother: United States, 1980-99; U.S. Census Bureau, *Statistical Abstract of the United States: 2000*, Table No. 85. Births to Teens, Unmarried Mothers, and Prenatal Care: 1985 to 1998, 69.

6. Roberto Suro, "For Women, Varied Reasons for Single Motherhood," *New York Times*, May 26, 1992, A12.

7. Immanuel Jakobovits; *Jewish Medical Ethics: A Comparative and Historical Study of the Jewish Religious Attitude to Medicine and Its Practice* (New York: Bloch, 1975), 266.

Chapter 2

Looking to Jewish Law and Beyond Regarding Questions of Sexuality and Procreation

From a Jewish perspective, there are two basic approaches to questions of sexuality and procreation.[1] First, the traditional approach accords paramount authority to Jewish texts as embodied in Jewish law (the *Halakhah* in Hebrew) as interpreted and applied by rabbis over the ages. The strictures of Jewish legal rules set limits on one's self-determination.

Second, the alternative approach focuses on our autonomy grounded in our Godlike nature and our partnership with the Eternal in perfecting the world. A considerable degree of decisional choice is implicit in the notion of being created in God's image. Looking to autonomy, not authority, rests on one's ability to make important decisions shaping one's life for oneself. Autonomy recognizes the ability of each of us to express one's own distinctive character -- one's own values, interests, and commitments. It permits each of us to lead our own lives, not subject to or being led along by a textual tradition, no matter how ancient, or by rabbinical authority.

As we will see, in addition to using divergent modes of reasoning and often yielding different conclusions, these two approaches differ as to who is the decision-maker. It is a rabbi under the first approach. It is each of us under the second approach.

In this chapter and throughout this book, I develop the nonlegalistic notions of Spiritual Judaism, an approach rooted in our autonomy. Spiritual Judaism focuses on: the One Divine Presence and the adherence to an ethical approach in our conduct with others. The prophet Micah expressed it so well more than two thousand five hundred years ago: "...[D]o justice, and ...love goodness, and ...walk humbly with your God" (Micah 6:8). As you strive to formulate your own approach to questions of sexuality and procreation, Spiritual Judaism guides and helps you exercise your autonomy, rooted in the virtues of love, compassion, and personal responsibility.

The Traditional, Text-Centered Approach

The traditional approach of Jewish ethics identifies precedents from biblical and rabbinic sources to arrive at Jewish principles and prescriptions applicable to a specific situation.[2] Thus, a qualified rabbi, as the ultimate decision-maker, first turns and looks to primary authorities, biblical and rabbinic.

Biblical texts include the Five Books of Moses, namely, Genesis, Exodus, Leviticus, Numbers, and Deuteronomy, known as the Torah (the teaching) or the Pentateuch, as well as the Prophets and various writings including the Psalms and Proverbs. These biblical texts were mainly written from the tenth to the fifth century B.C.E. The Torah is often referred to as the written law in contrast to the oral law representing the unwritten tradition interpreting and applying the Five Books of Moses.

Rabbinic Judaism was born out of the destruction of the Second Jewish Temple in Jerusalem some one thousand nine hundred years ago. For the next 500 to 900 years of the Common Era, up until the medieval period in Europe, a vast body of literature developed reflecting the creative ferment of the rabbis. The rabbinic literature consists of two parts, the Talmud and Midrash. The Talmud, mainly a legalistic discourse on all aspects of Jewish life, representing a reduction to writing of the oral law consisting of the opinions and teachings of numerous rabbis who analyzed and commented on the Torah as well as various nonlegal materials. Actually, there are two Talmuds, a Babylonian Talmud and a Jerusalem one. The former is the more complete and authoritative. The second part of the rabbinic literature, the Midrash, consists of written interpretations of Biblical passages using allegories, parables, analogies, and stories to illustrate the meaning of Torah verses.

Rabbinic sources from the medieval period include various commentaries and legal codes. These medieval rabbinic sources function today as statutory materials. The major medieval commentary on the

Hebrew Bible and the Talmud is that of Rabbi Shlomo ben Isaac, who lived from 1040 to 1105, and is known to us today as Rashi. Rashi's commentaries had an enormous influence on the development of Jewish law. The most illustrious Jew of the Middle Ages was Maimonides, a rabbi, philosopher, physician and ethicist, who lived from 1135 to 1204. His most famous work, the classic and authoritative multivolume, *Mishneh Torah*, represents a monumental compilation and systematization of biblical, Talmudic, and post-Talmudic law. The *Shulchan Aruch (The Code of Jewish Law)*, a unified code compiled by Rabbi Joseph Karo, who lived from 1488 to 1575, serves as the standard work of Jewish law. It has become Judaism's most authoritative legal code.

The major rabbinic literature of the last four centuries consists of responsa -- case law, if you will -- that comprise formal replies to legal questions about individual cases addressed to rabbinic scholars. Numerous rabbis have authored volumes and volumes of responsa over the centuries.

Previous responsa, in which rabbis applied concepts derived from the primary biblical and rabbinic authorities to specific problems, as well as various Jewish codes, such as constructed by Maimonides and Karo, serve as the precedents for traditionally oriented modern day rabbis. Although these responsa constitute precedents and describe the specific situation necessitating a decision, no two cases are exactly identical. The specifics of each case, of course, play a large role. The writings of more contemporary rabbis are also analyzed, thereby clarifying the purpose and meaning of earlier Jewish materials.

In grappling with a current situation, rabbis reason by analogy. They consider not only prior positions on a comparable issue and their reasoning, but they also apply principles derived from somewhat similar fact patterns deemed appropriate to the situation at hand.

Divergences in opinion arise from how one reads, stringently or leniently, prior precedents. Opinions also differ as to which sources are analogous to the situation under consideration and thus must be consulted. For some, precedents may be excluded from consideration as irrelevant; while for others the same precedents may be viewed as extremely relevant, if not controlling.

A text may be read literally or in accordance with the principle of purposive interpretation, that is, a text's "true" intent. Thus, giving rabbis the discretion to use the principle of purposive interpretation or other extra-legal factors, such as the Holocaust, often leads to a broader, more flexible construction of ancient texts. However, some traditionalists maintain that the sources must be read narrowly and mechanically in the search for one "correct" answer to questions seemingly offering a plurality of approaches.

In short, contemporary traditionalists sift through Jewish sources, both primary authorities and more modern precedents, in deciding into what category a specific situation falls. They then apply, by analogy, these primary authorities and precedents. Thus, even when authorities use the same methodology and read the same sources, they may read the texts differently and reach quite different conclusions. Sources are malleable. Rabbis can combine texts and manipulate sources to reach a desired conclusion. The personal predilections and values of different rabbis, as well as their theological assumptions, their implicit views about what it means to be human, moral considerations, the social-historical situation in which they find themselves, and their accompanying psychological state, play a role, often a key role, in the interpretative process.

No formal hierarchical structure exists for various contemporary, text-oriented rabbinic authorities. Even when one view surfaces as an authoritative consensus, laypeople may defer to their local rabbi, whose position is decisive.

For text-centered traditionalists, a rabbi (a *posek* in Hebrew), who has studied and mastered Jewish sources and is committed to "tradition," serves as the decision-maker. Rabbinical authority rests on Deuteronomy 17:8-13, where we read that each generation's questions about Jewish "law" must be addressed to that generation's judges, in modern terms, rabbis. Thus, a qualified rabbi, acting as a legal arbiter, makes the decision, taking the form of a *halakhic* directive, applying Jewish legal rules and principles, on an individualized, case-by-case basis.

A sensitive *posek* recognizes the gravity of any situation, especially those involving sexuality and procreation. This type of *posek* asks: does the pressure for leniency stem from a frivolous or a serious attitude and does it reflect an upright or a debased moral compass?

Yet, even in the hands of the most sensitive *posek*, the authoritarian decision-making structure of the *halakhic* system contravenes, for most of us, our notion about the desirability of decisions arrived at by responsible individuals. Many of us resist surrendering our autonomy to absolute authority. We want to be able to critique and question the teachings recorded by tradition. We sense that general and enduring conclusions to difficult questions are not always possible. We remain open to ambivalence in case of complex issues. There may be more than one "right" answer to a question, no "right" answer, or a choice from among a range of alternatives. Also, the problems presented by technological advances, considered in Chapters 6, 7, and 8, are unique to our time. Ancient texts, and centuries of precedents derived therefrom, often offer little or no guidance in resolving our quandaries.

A More Liberal, Text-Centered Approach

More liberal (Conservative, Reform, and Reconstructionist) rabbis and commentators apply a similar text-centered methodology in approaching ethical issues. The Conservative Movement has appointed its Committee on Jewish Law and Standards to provide interpretations of Jewish law. The Committee validates acceptable options on questions raised by individual rabbis. It also adopts standards binding on Conservative rabbis and synagogues. The Reform Movement's Central Conference of American Rabbis has created a Committee on Halakhic Inquiry and has published several books of Reform responsa, now available on-line at ccarnet.org, functioning in an advisory capacity for Reform Jews.

For these more liberal authorities, who are continuing to use a text-focused approach, a decision flows whenever possible from an interpretation of legal precedents derived from a variety of written Jewish sources. However, this decision is aided and enlightened by an assessment of ethical, historical, scientific, sociological, psychological and cross-cultural materials. Thus, these less literal commentators take into consideration the various meanings texts have as well as their historical and literary context.

For these liberal authorities, accommodation can occur within the Jewish legal system. Jewish law can undergo appropriate changes thereby preserving and enhancing the *Halakhah*. One of the most thoughtful of the modern interpreters of the Jewish tradition, Conservative Rabbi Elliot N. Dorff, Professor of Philosophy at the University of Judaism in Los Angeles, maintains, "We should retain a legal method with its inherent discipline, authority, continuity, coherence and educational utility in making our moral decisions and in teaching moral sensitivity, recognizing after all that the law must be continually tempered by its traditional and necessary interaction with morality."[3]

A text-centered approach, whether in the hands of traditional or liberal commentators, suffers from one fundamental weakness: the Jewish texts on which they rely were not Divinely revealed. They are human documents reflecting our spiritual creativity.

Questioning a Text-Centered Approach: The Jewish Texts Were Not Divinely Revealed

In going back to and using Jewish sources to guide us in making decisions regarding sexuality and procreation, it is important to realize that the Torah and subsequent rabbinic interpretations were not Divinely revealed. Jewish traditionalists, particularly Jewish fundamentalists, never want to discuss or face the virtually unanimous conclusion of scholars that the written law (the Torah) and oral law (the Talmud and later rabbinic interpretations) were not revealed by God to Moses on Mount Sinai. Rather, the Torah achieved its familiar shape and dimensions by various human hands as the result of an evolutionary process. The books of the Hebrew Bible were written at different times by various writers. Thus, the Torah, the Talmud, and the Jewish law are the products of human beings striving to understand God.

For the past four hundred years, critical biblical analysis has found the following: numerous duplications and repetitions in the Pentateuch; variations in the name of the Divine; and a broad diversity of language and style and contrasting viewpoints in the text of the Torah.

The discovery of duplication extending across a considerable body of the Five Books of Moses, starting with two stories of creation in the Book of Genesis (Genesis 1:1-2:3 and Genesis 2:4-24), led to the claim, going back to the seventeenth century, that the Torah had its own history of composition. Specifically, it had been compiled from a number of documents.

Variations in God's Hebrew names, Elohim and Yahweh, used in the Pentateuch, provide further evidence of different sources.

Three parallel sources (E (for Elohist), J (for Jahwist), and P (for Priestly)) and the addition of a fourth source, D (for Deuteronomist), gave rise to what has become known as the four-source documentary hypothesis.[4] Although scholars vigorously dispute the dates and the exact nature of the four sources, the view that four sources underpin the Torah, as we know it today, holds the field.

Thus, modern, critical biblical scholarship has undermined the traditional understanding of the Torah as divinely revealed by God to Moses. Biblical scholarship called (and continues to call) into question two bedrock doctrines: first, a literal revelation and second, the covenant between God and the Jews, as the Eternal's chosen people, on which much of traditional Jewish law is based.

The authors of the Five Books of Moses wove together various source material, a collection of traditions, semi-historical stories, and sheer fantasies, originating at different times in ancient Israel and neighboring cultures. The Pentateuch, as a human work, was authored by many humans and was subject to editing and re-editing over long periods of time.

Although the Torah has special significance for the Jewish people, these writings are not absolutely or ultimately true or a superior form of Divine revelation. Rather, I suggest that we view the Torah as a record of the spiritual life of the Jewish people and as evidence of their quest to know God.

Viewing the Pentateuch as a product of humans, an unverifiable mix of myth and historical fantasy, not the commanding voice of God, we in the twenty-first century need not accept as binding, definitive authority for our lives either the Torah or the two thousand years of rabbinic interpretations. We need not live in accordance with all of the supposed 613 Divine commands designed to take Jews through all aspects of living and dying. Biblical literalism represents a form of idolatry -- the worship of a humanly written text.

We are thus led to a most difficult and challenging inquiry. If not rabbinic authority and the *Halakhah*, the two pilars that have provided a sense of security and order, a feeling of tradition and continuity, then what is to guide Jews in the twenty-first century?

Refashioning Our Approach to Sexuality and Reproductive Decision-Making within the Framework of Spiritual Judaism

I write this book within the context of Judaism -- what I call Spiritual Judaism, an ethical, virtue-oriented, monotheistic system, based on Jewish sources, ancient and modern -- beyond the minutiae of the traditional legalistic rules designed to govern everyday living.[5]

Striving to fill our spiritual hunger as well as to offer meaning to our existence and a connection with the Transcendent, Spiritual Judaism represents a nondogmatic, nonlegalistic path, not relying on Jewish law. It is a way of thought, feeling, and belief, not centered on rules, about God and the spiritual depths of our being. It welcomes all people who seek to attune to the Divine Presence in their lives. It offers the vision of a personal, intimate experience of the Eternal as the source of health, joy, love, abundance, and wholeness. It is designed to help us meet not only the ups and downs of daily life, but also the crises we all face in living and dying.

Spiritual Judaism focuses on practical spirituality, spiritual living that is liberating, promoting and unfolding our inner human possibilities, helping each of us realize his or her highest self and that of others around us. In providing norms for "sound" behavior, Spiritual Judaism rests on three pillars: first, our concept of God; second, our relationship to the Eternal; third, our relationship to one another and the world based on certain virtues, notably, love, compassion, and personal responsibility.

Our Concept of God: The Approach of Spiritual Judaism

From the perspective of Spiritual Judaism, God, as the ground of all potentiality, manifests an all-sustaining care for humanity. In the words of the Psalmist, the Eternal is our compassionate, loving parent (Psalm 103:13), Who desires our happiness. God is the loving Helper, Who cares for everyone and Who is ever-present as our companion and guide.

The Holy One has a beneficent interest in every human being. God watches over and is interested in and concerned with everyone the Eternal has created. Each of us matters to God; each of us is precious to the Holy One, Who seeks to bring forth goodness, beauty and adventure in our lives.

For spiritually-oriented Judaism, the main attribute of God is unconditional love. The Eternal serves as the source of all loves, compassion, and goodness in the world. As the Psalmist (Psalm 116:5) reminds us, "...[O]ur God is compassionate."

Through love God called, and continues to call, everyone into existence. The Eternal's love sustains the world. The Holy One offers hope and happiness for everyone.

The Hebrew prophets saw the Holy One's goodness and mercy. According to Ezekiel (Ezekiel 18:23): "Have I any pleasure at all that the wicked should die?" says God. "It is rather they should repent and live."

Hosea (14:5) views the Divine mercy as follows: "I will heal them freely, generously will I take them back in love; for My anger is turned away from them."

When God, the Eternal Goodness, answers the Psalmist's prayers, the latter's difficulties and despair vanish. In Psalm 30:12, we read that the Holy One is the source of goodness and the fountain of all joy. The Psalmist writes: "You did turn for me my mourning into dancing, You did loose my sackcloth and girded me with gladness...." In Psalm 106:1, the Psalmist declares: "God is good, the Eternal's love and mercy endure forever."

Today, we continue to walk in the light of God's love, peace and joy. Through our awareness of God's presence and guidance, we feel peaceful within -- a calmness enabling us to face any situation with confidence and creativity.

Our Relationship to God: The Approach of Spiritual Judaism

Humanity serves as God's partner in the ongoing act of shaping, improving, perfecting and making whole an unfinished world. The hoped for Messianic Era will usher in not only a time of peace and justice for everyone but also a period of human self-realization. At the beginning of the Messianic Era, all humans will be treated like the Divine and will become Godlike.

The notion that in perfecting the world the Eternal requires human help rests on two fundamental Jewish values: first, humans are created by God in the Divine image; and second, humans are autonomous. We read in the Book of Genesis: "God created humans in the Eternal's image...." (1:26-27; 5:1, 9:6). Because humans are created in the image of God, we partake, on a small and often rather imperfect scale, of the Eternal's attributes. Specifically, we possess the Divine attribute of free will and choice (Genesis 3:5, Deuteronomy 30:19).

The Holy One's power is expressed in extending itself to us so that we may exercise our freedom of choice and creativity. We are not puppets. We can bring great good to others and one's self, or we may cause injury and evil. We are free to respond to the circumstances in which we find ourselves, how we perceive and experience them, how we interpret what happens to us, and the choices we make in life.

Through our mind and our way of thinking we were socialized into, we envision the world we see. We can create our opportunities; we can change our destiny. Thus, our future is not made in the womb or at birth, rather it is within our will to create it, subject, of course, to certain boundaries, such as our genes and our environment.

Because of our Godlike nature as well as our freedom and our capacity for thoughtfulness, we are called to participate in the ongoing process of improving the world and completing the work of creation. This is called repairing or healing the world (in Hebrew, *Tikkun Olam*). We can co-create with God to loving beauty to ourselves, our families, and the world. Each generation does what it can; then the task is taken up by the next generation and so on, until the perfection ushered in by the Messianic Era will be realized. Each of us has a part to play in the human chain designed to shape the world and remove "imperfections."

Human autonomy reflects the dignity of both God and the Eternal's worldly creation. Humans become more and more Godlike as a result of their ongoing partnership with the Holy One. As Reform Rabbi Peter Knobel writes, "In such a loving relationship human freedom is not overwhelmed by divine will. Such a relationship is characterized by mutuality and respect for the integrity of each partner."[6]

The ability of humans to make and act on their decisions about their lives rests on the freedom of choice the Holy One gives each of us. By empowering humans, the Eternal makes us more Godlike. Each of us is free to make choices throughout life, recognizing the uniqueness of each of us and that God is the source of our freedom and creativity. The Holy One wants each of us to say "yes" to the freedom that lies before us. One size -- one rule -- does not fit all.

This approach to decision-making impacts on who serves as the ultimate decision-maker. A rabbi may provide input into the decision-making process. Acting as an advisor, a rabbi may provide information drawn from the Jewish tradition. Within the framework of Spiritual Judaism, each one of us will, however, ultimately make our own decisions. Thus, because of humanity's relationship with God, this alternative approach accords a key role to each one's autonomy.

In making decisions regarding sexuality and procreation in light of our autonomy, what sort of actions should one pursue as a Jew? The virtues underpinning Spiritual Judaism, namely, love, compassion and personal responsibility provide a framework for the exercise of our autonomy, preventing our descent in narcissism and amorality.

Love, Compassion, and Personal Responsibility: Virtues as Keys to Spiritual Judaism

It is important for Judaism to emphasize our ethical responsibilities as a corollary of and as a guide for our autonomy and the exercise of our free will. We need to conduct ourselves ethically even in the most complex and difficult moments of our lives, for instance, choices regarding sexuality and procreation.

The ethics of Judaism, which have become part of our Western heritage, are as necessary and as relevant in the twenty-first century as when first delivered. Judaism should mold our daily conduct, influence our relations with others, and make us more honest with the world, zealous for justice and righteousness.

In addition to being just and truthful in all of our relationships, we should strive to do the "good and beautiful" by living in a loving and compassionate manner. We need to act with love and compassion, to speak kind words, and accept responsibility for our conduct and speech.

Love. Spiritual Judaism rests on bringing us into an ideal relationship with others. We are here to express love which draws people together. We are here to be the presence of love.

We are endowed with the power of love by God. Love is inherent in all of us. It is intertwined with the roots of our being. God's continuous Presence within us causes us to love. Our love flows from the Holy One's love. As noted earlier in this chapter, the Eternal is love, specifically, unconditional love. Everyone has within him or her a spark of God's love.

Love represents an all-embracing, transcendent expression of the Divine. The highest and noblest manner in which love finds expression is in our love for others.

In the very personal context of this book -- sexuality and procreation -- love is a powerful, living emotion that must be nurtured. This type of love partakes of emotional, spiritual, and physical aspects. Intimate love involves passionate feelings and an intense closeness.

The Biblical Song of Songs, containing songs of love and nature, of courtship and marriage, reveals the emotional and spiritual aspects of love and glorifies its physical aspects. Declaring the beauty and power of intimate love, the author of the Song of Songs (8:6-7) tells us: "Let me be a seal upon your heart, like a seal upon your hand. For love is as powerful as death, passion is as strong as death itself; its darts are darts of fire, a blazing flame. Vast floods cannot quench love, nor rivers drown it. If a man offered all his wealth for love, he would be laughed to scorn."

In intimate human relationships, the greatest happiness is attained through love. Each of us seeks a relationship in which another lavishes love on you and in turn you have someone to lavish love upon. When you choose another to be the object of your love, both must willingly feed that love so that daily it grows stronger.

God instilled love in our hearts to enable us to establish a binding relationship with another and in order to augment the happiness of both. If there is love, there is happiness. When one expresses love, one creates happiness.

We long for love, and especially for companionship and intimacy, that makes life complete. We seek someone to whom we can impart our deepest feelings, someone to entrust with our innermost thoughts, someone with whom to share our joys and who will participate in our woes.

Love means a willingness to sacrifice oneself for another, without any ulterior motive. Those truly in love manifest not only mutual respect and concern but also desire to make the other happy at any cost. Each relates his or her desires to the other's concerns and needs.

Love and sex are interrelated. Judaism sets forth a healthy, affirmative attitude toward sex. The emotion of love finds expression in a couple's physical desire for each other. In the Song of Songs (7:7, 11), one lover declares: "How fair and how pleasant are you, O love, for delights." His beloved responds, "I am my beloved's, and his desire is for me." The love of two individuals is inseparable from the physical sexual experience, representing the expression of the highest level of intimacy. Sexual expression makes the bond between two humans even more profound.

Thus, love in its emotional, spiritual, and physical fullness calls us to a enduring life of ever deeper meaning and purpose. Sharing true love makes life worthwhile. To love and in return be loved makes both fully human, fully what they might be.

Compassion. As outsiders to two people in an intimate, loving relationship, each of us needs to manifest compassion -- a matter of feeling with and for others. It rests on the notion that your heart will break and that you cannot bear what another is (or what two others are) going through. By acknowledging and opening your heart to others, helping them, relieving their suffering, and encouraging their love, you feel your unity with others.

Being compassionate, according to the Jewish tradition, enables us to be more Godlike. According to the rabbis: "Resemble the Holy One, please! Just as God is merciful and gracious, you too must be merciful and gracious." (Mekhilta de Rabbi Ishmael on Exodus 15:2). As Rabbi Joshua interpreted in Rabbi Levi's name: "God imparts the Divine attribute of mercy to humans so that they may be merciful like the Holy One." (Midrash Genesis Rabbah Parashah 33:3). As God is merciful and compassionate, so too you should be merciful and compassionate towards others.

Personal Responsibility. As co-creators with God, we must assume responsibility for all our words and our deeds. Responsibility is an important virtue based on our active partnership with God in improving the world (Babylonian Talmud Shabbat 119b). Self-mastery represents the ability to follow one's better judgment, despite the tug of other desires, personal or social.

Humanity's role in perfecting the world rests on our autonomy, specifically, our free will -- our freedom to live the life we choose -- and as a corollary, to accept our responsibility. We are morally free. We have the ability to distinguish right from wrong and to act and to speak accordingly. The author of the Book of Deuteronomy tells us (30:15): "I [God] place before you a choice between good and evil, between life and death." Each of us possesses the rather perilous privilege of freedom of choice, perhaps God's greatest gift to us, and with it we must assume full responsibility for our decisions. Making humanity a partner with God in the ongoing work of creation represents our greatest privilege and our gravest responsibility. Using our constructive will, each of us is responsible for who we are and what we make our life.

In accepting the challenge of personal responsibility, we must strive to foresee the consequences of our conduct and speech and not engage in acts or talk that we later will regret. Thus, we ought to restrain the harmful tendencies in our being and allow them no expression. Each of us daily needs to struggle against noxious tendencies within one's self and strive to triumph over them. One who is master of oneself is not impetuous or impulsive. In charting a course of conduct in the most intricate of situations, we must seek to distinguish between the harmful and the beneficial, the detrimental and the wholesome, allurement and benefit to ourself, to another, and to society.

The area of sexuality, encompassing both the search for a partner and the act itself, raises questions of personal vulnerability. Each partner needs to remember that sexual involvement may be of great, if not overwhelming, significance to the other. Responsibility entails consideration for the feelings of, and the life consequences for, one's partner.

It is up to each of us to decide what role we want Jewish sources and virtues, as illuminated by Spiritual Judaism, to play in our sexuality and procreation decision-making process. I want to encourage each of us to take responsibility for our life and to think for ourselves, making decisions for ourselves, in an informed, thoughtful, and responsible manner, enlightened by, but not shackled by, the Jewish tradition and its modern interpreters. As we turn and consider sexuality, marriage, and the family in the next chapter, let us consider how we can make our decisions regarding pre-marital sex in a manner consonant with our autonomy, as bounded by the virtues underpinning Spiritual Judaism, namely, love, compassion, and personal responsibility.

Notes

1. I have drawn on David H. Ellenson, "How to Draw Guidance From a Heritage: Jewish Approaches of Mortal Choices," in *A Time to be Born and a Time a Die: The Ethics of Choice*, ed. Barry S. Kogan (New York: Aldine de Gruyter, 1991), 219-232, reprinted in *Contemporary Jewish Ethics and Morality: A Reader*, eds. Elliot N. Dorff and Louis E. Newman (New York: Oxford, 1995), 129-139; Eugene B. Borowitz, *Choices in Modern Jewish Thought: A Partisan Guide* (New York: Behrman, 1983), 243-289.
2. David M. Feldman, *Birth Control in Jewish Law: Marital Relations, Contraception, and Abortion As Set Forth in the Classic Texts of Jewish Law* (Northvale, NJ: Jason Aronson, 1998), 3-18 provides a useful overview of the structure of Jewish law.
3. Elliot N. Dorff, *Matters of Life and Death: A Jewish Approach to Modern Medical Ethics* (Philadelphia: Jewish Publication Society, 1998), 416.
4. The scholarly insights on documentary hypothesis are brilliantly summarized by Richard Elliott Friedman in *Who Wrote the Bible*? (Englewood Cliffs, NJ: Prentice Hall, 1987).
5. See generally, Lewis D. Solomon, *Jewish Spirituality: Revitalizing Judaism for the Twenty-First Century* (Northvale, NJ: Jason Aronson, 2000).
6. Peter Knobel, "Suicide, Assisted Suicide, Active Euthanasia: A Halakhic Inquiry," in *Death and Euthanasia in Jewish Law: Essays and Responsa*, eds.

Walter Jacobs and Moshe Zemer (Pittsburgh: Rodef Shalom Press, 1995), 31.

Chapter 3

Marriage, Family, and Pre-Marital Sex in the Jewish Tradition

Sex today is not seen as special or sacred. There is an erosion of awe in sexual matters. We accept what now seems inevitable -- everyone does it. Yet, sex should be more than impulse, instinct, and instant gratification.

Judaism seeks to introduce into human earthly existence an element of holiness (*Kiddushah* in Hebrew) (Leviticus 11:44-45, 19:2, 20:7, 26). Sex should be imbued with a sense of holiness and sanctity -- to be set apart and elevated to a higher level of living. The mysterious power of sexuality should open us to the Transcendent Realm.

Teaching that we need to handle the sexual impulse with care, the Jewish tradition has sought to channel sexual activity -- the most powerful force in human experience -- into marriage, based on the enduring fidelity of a husband and a wife. For Jews in the twenty-first century, norms restricting sexual intimacy to heterosexual marriage may require modification. Changing mores and social realities necessitate a searching reexamination. But where to draw the line and how to deal with the contemporary climate of sexual permissiveness and licentiousness?

After examining marriage, the family, and sex in the Jewish tradition, this chapter focuses on the prohibition against pre-marital sex in the context of modern American trends. Viewing sex as more than a biological function, we need to delineate where sexuality and spirituality come together. While heterosexual marriage is no longer the only key to the sexual ideal, the relationship of heterosexual couples engaged to be married in the near future is sufficiently imbued with the sacred to permit sexual intercourse. While encouraging cohabitating couples to marry,

Judaism ought to create a ceremony with provisional vows for those couples living together, thereby facilitating their journey of sexuality and spirituality.

Marriage and Family in the Jewish Tradition

For most of us, marriage is the most important thing that occurs in our lives. It is a unique personal relationship between husband and wife based on the exclusiveness of sexual relations. A faithful, monogamous marriage represents the transformation of physical love into biological and social continuity transcending the two participants. Ratifying the love between two individuals, it is an investment in the future of the family unit and the social order. The importance of marriage, as a public institution creating social stability, cannot be underestimated.

Marriage represents a permanent, intense bond and the unrelenting involvement of two people to live out the rest of their lives together. It involves a public statement of both partners' commitment to each other, before their families, friends, and society at large, that they will undertake a lifelong partnership in which they pledge fidelity to the serious responsibilities of family life.

Loving human relationships, sanctified by a marriage ceremony, are sources of holiness. The Jewish marriage ceremony is called in Hebrew, *Kiddushin*, and connotes a setting apart, sexual exclusiveness. Jewish homes were and continue to be founded on the sexual faithfulness of husband and wife to each other.

The holiness of marriage signifies a level of joyful and sacred devotion, tenderness, and interdependence. A married couple strives to evidence constantly a unique concern for one another, assuming responsibility and care for each other, helping each other through life's joys and sorrows. One heart makes room for another heart, one mind encompasses another mind. One soul invites another into its essence. Both lives become enhanced, both hearts become enriched, both personalities gain in depth. Throughout a lifelong marriage, new efforts must be made daily to gain more love, to achieve ever deeper devotion, and to realize the ideal of a more perfect union.

Marriage provides a couple the companionship of each other, it offers the opportunity for procreation, and it enables a husband and a wife to share in the rearing and educating of children.[1] The Book of Genesis presents a view of marriage as the means to satisfy and fulfill the basic human need for companionship and human intimacy, relating that God announced: "It is not good for a person to be alone" (Genesis 2:18). Therefore, "a man leaves his father and his mother and clings to his wife, so that they become one flesh" (Genesis 2:24). After the legendary Adam experienced the pain of loneliness, he more readily appreciated the need for companionship.

Each human, the Jewish sages tell us, best lives a meaningful and purposeful existence when linked in a stable, intimate relationship with a mate. As the Babylonian Talmud (Yevamoth 62b) teaches: A Jew without a spouse lives without joy, blessing or happiness.

Husband and wife become loving friends sharing what is going on in their lives and what they feel and think. Marriage fosters mutual affection, care, trust, sacrifice, and support, encouraging personal growth -- intellectual, moral, and spiritual. It brings two individuals together in a creative partnership for meeting life's inevitable difficulties. Shared anxieties and sorrows are divided; hopes and joys are multiplied.

Another aspect to the holiness of marriage is that is enables a couple to procreate -- to create and nurture new life in the image of God, thereby transcending their own existence. It provides a context for nourishing children -- morally, emotionally, educationally, and physically. Childbearing and child rearing, within the context of heterosexual marriage, represent traditional Jewish values. According to the author of the Book of Genesis (1:28, 9:1, 7), God said: "Be fruitful and multiply." Having children is regarded as a blessing. Concern for the preservation and continuation of the Jewish people has underpinned the Jewish commitment to this value. Children denote continuity. Marriage institutionalizes the desire for continuing the chain of human existence and spurs it on.

Through children, parents renew their sense of purpose and joy in life. Parenting is a transformative experience. There is a special fulfillment and joy in raising children. While it entails extraordinary effort and some measure of anxiety, on the whole, most parents would likely say that the benefits of parenthood predominate. The gains far outweigh whatever they might otherwise have done -- work or leisure -- had they not had children.

Marriage also enables parents to educate their children in the Jewish tradition. Family education, preparing the next generation of children to be Jews, is a vital key to the continuity of Judaism. As the legendary patriarch Abraham was charged with instructing his children (Genesis 18:19), we are also told to do so (Deuteronomy 6:4-9, 11:13-21). As the locus of child rearing and teaching by example, the family serves as the vehicle for the transmission of Jewish identity, values and practices. Through marriage and the family we link ourselves to the Jewish past, present, and future.

The family is the fundamental social unit. Couples assume family responsibilities not only for their own pleasure and growth, as well as that of their children, but also for the continuity of the Jewish people. The traditional view is that the family serves as the primary foundation of the Jewish people -- a bulwark and refuge against outside forces.

The family is perhaps Judaism's most important religious institution and one of its major contributions to human civilization. Most of us become fully human and achieve personal maturity -- emotionally and psychologically -- only in marriage, with its commitment, trust and involvement in building and providing spiritual, emotional, and material security and stability for one's family.

Sex in the Jewish Tradition

Sexual intercourse is the greatest and most powerful of human experiences, the most intimate human act, generating deep feelings, uniting body and spirit. Sex represents a supreme physical, emotional and spiritual experience. The proximity of another human being is the ultimate act of a personal relationship. It is perhaps the most extraordinary act we are regularly capable of. Male-female sex also has a unique aspect -- it may create new life.

Sexual intercourse is a source of physical enjoyment -- physical pleasure and release. The positive physical sensations generated by sex make a couple feel good about each other and this strengthens their bond. It is a form of communication and a means to express love for another. We desire the physical and emotional fulfillment sex provides.

Modern American culture portrays the body, and particularly our sexuality, as a pleasure-generating machine. We are told to use one's body for as much pleasure as possible.

Although regarded as a normal, healthy human activity, sex, however, occupies a different position in the Jewish tradition. Each of us is an integrated whole of body-mind-spirit -- with each part continuously impacting on the other aspects. Thus, sexual intimacy is physical, emotional and spiritual. Sex represents an intense way each of us relates to another physically, emotionally, and spiritually.

Sex enables us to reach out beyond ourselves and attain pleasure. But, sex is more than using another for one's physical release. Sexual intercourse ought not to be mere mechanical or casual physical activity designed for personal pleasure. It needs to express a couple's emotions -- their mutual love and respect for each other. The emotional component of sexual intimacy is essential to maintaining a person's well being.

Sexual desire and intercourse are part of the process of spiritual unfolding. Sex enables us to achieve wholeness and holiness. Depending on the state of a couple's consciousness and intentionality during their lovemaking, sex enables humans to come closer to God. Sex is more than physical release or a result of animal-like lust; it enables the expression of the Divine image within each of us. In short, sex is a positive, sacred means to promote human relationships and spiritual self-development. We must use our sexual attributes to fulfill their Divine potential.

Although seeing sex in a positive light, the Jewish tradition rests on a stern morality. The Talmud (Babylonian Sanhedrin 75a, Jerusalem Abodah Zarah 2:2) tells the story of an unmarried man who sought permission for intercourse with an unmarried woman with whom he was infatuated. As a result of his desire, he became physically ill and sought an unusual remedy. The physicians with whom he consulted agreed to the sexual intercourse, but the rabbis objected, stating: "No, let him rather die." The physicians suggested that the woman simply stand naked before him, but the rabbis objected to this as well. The physicians asked that he simply be permitted to speak to her from behind a fence, but the rabbis also rejected this and again said it was preferable that he die.[2]

For the past fifteen hundred years the Jewish tradition has sought to maintain this high standard of morality. The rabbis sought to avoid contact between men and women to prevent sexual intercourse outside marriage. They highly esteemed virginity. Because the male is constructed to be visually aroused, they devised stringent standards for female modesty. For instance, the Babylonian Talmud tells us to drive immodest women from the city, as happened to Homah, a widow, who exhibited her arm in a rabbinic court (Ketubah 65a). Through their moral interdictions, Jewish sages sought to uphold the dignity of women and prevent their being regarded as mere objects of sexual conquest.

By custom, women in ancient times, especially unmarried women, were expected to spend most of their time at home. When they ventured out, they dressed so no part of their body showed. They wore some type of head covering so that hair, deemed sexually arousing, could not be seen in public (Babylonian Talmud Ketubah 72a, Berakoth 24a).[3]

Traditionalists made (and continue to make) every effort to keep the sexes apart. Women are segregated at public functions, at worship, and at weddings.[4]

Men were forbidden to be alone with women in a room or visit them socially, without chaperons. The need to supervise the private meetings of men and women led to chaperoning boys after age nine and girls on reaching age three (Babylonian Talmud Berakoth 24a). A man could not touch -- not even shake hands in greeting -- or kiss any woman, including adult relatives, except his wife. The rabbis propounded numerous rules, such as not permitting a man to walk behind a woman (Babylonian Talmud Berakoth 61a, Erubin 18b, Kiddushin 81a), or to gaze or directly stare at her, even at her little finger (Babylonian Talmud Avodah Zarah 20a-b, Berakoth 24a, Erubin 18b, Nedarim 20a). Even hearing a woman's voice, one sage remarked, "stimulates sexual longing" (Babylonian Talmud Berakoth 24a).[5]

Today, even progressive Jews ought to treat their bodies with modesty. Although women now dress according to contemporary styles, clothes ought not to accentuate sexually arousing body parts. Sexually suggestive or revealing clothes are inappropriate. Crass or profane sexual language indicates discomfort with one's body, disrespect for others, and for the Divine nature of all.

Beyond standards of attire and language, both traditionalists and more liberal authorities today continue to offer a firm "no" to pre-marital sex.[6] Sex is reserved for marriage, the most all-encompassing of human relationships. The Conservative Movement's rabbinical association, The Rabbinical Assembly asserts, "Judaism posits marriage as the appropriate context for sexual intercourse."[7] A recent Reform responsum concludes, "We will not be party to steps that undermine Jewish marriage or that diminish ambitions for sanctity in Jewish lives.... [S]exuality *within* marriage is our ideal."[8] Living in a society constantly challenging this ideal, how should Judaism respond?

Modern Trends in Pre-Marital Sexuality

Pre-marital sex is rampant in contemporary America. For example, in 1999, 52% of high school males have had sex (down from 57% in 1991). Forty-eight percent of high school females in 1999 were reported as sexually experienced (as opposed to 51% in 1991). By senior year in college, men and women who are sexually experienced reaches closer to 85-90%.[9] Nearly all men and women have had sexual intercourse prior to their engagement, let alone their marriage. Individuals simply do not limit sexual intercourse to marriage.

Over the past forty years, sexual permissiveness has represented a revolt against moral standards. The search for self-gratification and the notion that "anything goes" arose from various factors, including a desire to flout traditional conventions, a rebellion against repressed, uptight attitudes, and the quest to live passionately and for the moment. Technology -- effective contraception and condoms -- have, to a large measure, surmounted the old fears of unwanted pregnancies and sexually transmitted diseases, allowing women to decide if and when they want to reproduce.

Our modern, highly sensuous culture, notably its erotic advertising, music, movies, videos and television, continually bombards us with sexual come-ons, seeking to harness the forces of sexuality to increase our consumption. Explicit sex is everywhere. We are overwhelmed with sexual messages.

Young adults face powerful sexual urges and intense peer demands. There are enormous pressures particularly on young women in the contemporary American culture, from their mid-teens into their 20s, to "put out" or be excluded from their social circle by sexual rites that arbitrate relations between sexes in high schools and on college campuses. Shifts in values have also made the loss of virginity of far less importance. Rather, it has become important to lose one's virginity.

The Jewish Tradition's Prohibition of Pre-Marital Sex

The Hebrew Bible contains no direct rule explicitly prohibiting pre-marital or extra-marital male-female sexual intercourse, with the exception of rape, prostitution, adultery, and incest. In fact, we are told Ruth climbed into the bed of the man she hoped to marry (Book of Ruth

3:7).

Although female virginity was prized in ancient times, a nonvirgin could marry without stigma. Seducing a virgin meant financial loss to her family and likely minimized her ability to find a husband. The seducer was forced to marry her. If her father refused to give her to him, the seducer had to pay a fine (Exodus 22:15-16). A man who wanted to marry a nonvirgin could freely do so.

Despite numerous lapses, the Jewish community looked to marriage as the only place for sexual relations (Babylonian Talmud Sanhedrin 76a). The Jewish tradition sought to regulate the sexual behavior of fertile heterosexuals because of their possibility of conception.

Jewish custom called for arranged teen-age marriages. Adults became engaged at an early age, women at 16 and men at 18. Pre-marital sexual intercourse by unengaged Jewish adults was a common problem in ancient and medieval times.

The best available evidence indicates a hardening of the Jewish position in the late Middle Ages leading to requiring a marriage contract and rites for sexual relations.[10] Judaism became committed, legally and morally, to sexual intercourse only in marriage.

In the twelfth century, Maimonides, in formulating his legal code, prohibited pre-marital sexual intercourse. Maimonides used the prohibition against cult (or religious) prostitutes, a type of free, no strings attached sexual encounter (Deuteronomy 23:18) as the basis for barring all unmarried sexual intercourse.[11] Viewing pre-marital sex as tantamount to prostitution, he saw the marriage bond as the Jewish tradition's advance over the laws and arrangements of other primitive societies. Marriage prevented the trivialization of sex -- the ultimate human relationship. He many have also reasoned that if people had indiscriminate intercourse children of unknown parentage would abound. These offspring might one day marry one another. Incest could become far more common.

Jewish mysticism, which also emerged in the late Middle Ages, with the publication toward the end of the thirteenth century of the *Zohar* (*The Book of Splendor*), strengthened traditional Jewish sexual standards, particularly the prohibition against sex outside marriage. In the mystical tradition, "properly intended," that is, marital, sexual intercourse impacts on and directs the Divine Sefirot -- the ten inner workings of the Divine -- to achieve full harmony and integration among themselves. The union of God, represented by the joining of the male and female aspects of the Sefirot, serves as the prototype for loving, male-female sexual unions.

When human intercourse is holy, God is unified. Sex for married couples makes possible their linkage to the Divine. The husband-wife relationship facilitates completion of the circuit of Divine electricity, charging all life. Proper intercourse, especially for procreation, causes the Divine energies to flow back to humans. Conversely, when sexual activity is immoral or improper, God's inner dynamic is disrupted. Illicit intercourse gravely damages the Divine Sefirot.[12]

Despite the efforts to channel sexuality into marriage, pre-marital sexual intercourse takes place: as casual sex with little or no love or commitment; in a loving, committed relationship; or after an engagement. How does (and how should) Judaism view sex in each of these contexts, particularly in light of contemporary realities?

Casual Sex

Although the Talmud mentions that a traveling sage on reaching a village would ask the local single women: "Who wishes to be with me privately for the day?" (Babylonian Yoma 18b), the Jewish tradition stands against casual sexual relations for mere physical pleasure, with little or no love or commitment.

Today, a "pick up" occurs when men and women meet in some common territory, like a bar. They "hook-up," seeking instant sexual gratification. Reducing any inhibitions with liquor, they exchange some congenial banter and pleasant talk and then agree to engage in sexual intercourse, without any pretense of deep intimacy. The consenting parties typically do not know each other well or for a longtime. They are not greatly attached to each other or love one another. They act on transient feelings of pleasure ("It makes me happy") without permanent commitment. Their feelings are as insubstantial as cotton candy.

Seeing sexual intercourse as the single most pleasurable thing most people can do, striving to get rid of their sexual frustrations, and propelled by modern contraception and cures for most venereal diseases (but not AIDS), casual sex enables two individuals to obtain the physical sexual fulfillment they seek and enjoy. Freely agreeing to participate, without any type of coercion or barter, they relate to each other on a biological level, mutually sharing the pleasure of intercourse for its own sake. Aiming at self-fulfillment, without any spiritual or relational assumptions, they strive to overcome the repression of their sexuality, seeing the other as a sexual channel, a means to achieve one's sexual pleasure, claiming a need for sexual expression, even without commitment, in order to be healthy persons.

However, sex for mere physical pleasure or release degrades us. If a man and a woman are not in love, their intercourse devalues the unique emotional and spiritual meaning of the sexual experience. Unprotected casual sex poses health risks, notably AIDS, presently incurable and lethal, and various strains of venereal diseases currently resistant to available drugs. Thus, sexual intercourse should be reserved for "special," not casual, relationships. It may sound "old fashioned" but Spiritual Judaism agrees with one Reform responsum that concludes: "We should do whatever we can to discourage casual sexual relations."[13]

Sex in an Ongoing, Loving Relationship

Promiscuity undermines the stable, enduring relationship needed for maintaining the social order and for rearing children. Today, many individuals marry years after puberty. As individuals pursue educational or other goals, sexually significant years pass by. Beyond mere mutual consent, a man and a woman, two unmarried adults, may engage in sex in the context of an ongoing, loving relationship. There are nearly 4.5 million unmarried, heterosexual couples in the United States who live together. Cohabitation accounts for 7% of all heterosexual couples, up from 1.1% in 1960, a virtually unprecedented lifestyle shift.[14] The percentage of first marriages preceded by cohabitation rose from roughly 10% for those marrying between 1965 and 1974 to over 50% for those marrying between 1990 and 1994. The percentage is even higher for remarriages.[15] While cohabitation before marriage has become the norm, long-term relationships are is still rare. Most cohabit for a few years and either get married or split up.

Those "in a relationship," characterized by loving concern, express respect and care for their partner in words, feelings, and actions. They think of the other's needs and desires and attend to them. They consider what the other is and will be, recognizing the other's potential and striving to help the other achieve it. One's beloved brings unparalleled delight and inner joy. Love fulfills and exalts. The sharing of love makes life worthwhile.

For those in an ongoing relationship, where each is concerned about the other's well-being, sex not only provides physical release but also companionship and intimacy. Their sexual expression represents another dimension of their love, not their infatuation. They express their love in their sexual relations. Through lovemaking, they find personal fulfillment and joy. This type of relationship is not causal, exploitative, or abusive.

In the Jewish tradition, there is a long-standing exception to the prohibition of pre-marital sexual intercourse. In biblical times, taking a *pilegesh* (a half wife) was permitted and practiced.[16] The legendary Abraham had Hagar (Genesis 16:3); while Jacob had Bilhah (Genesis 30:4) and Zilpah (Genesis 30:9). The judges (Judges 8:31, 19:1, 20:4) followed a similar pattern as well as monarchs, like Saul, David, and Solomon (II Samuel 3:7, 5:13, 15:16, 16:21, 19:6, 20:3, I Kings 11:3). In this semi-marital arrangement, a *pilegesh*, although lacking a marriage contract and a wedding ceremony, had a socially recognized familial relationship to a man. A *pilegesh*, a consenting, unmarried woman, and a man, married or unmarried, lived together in his home. She had sexual relations exclusively with one man although they were not married. This practice surfaced again in medieval Spain, where Jews lived in close proximity to Christians and Moslems. Extra-marital liaisons among the non-Jewish nobility and the affluent were common. Jews followed suit.

Although less strict authorities found a basis for these extra-marital relationships in the biblical institution of the *pilegesh*, by the late twelfth century Maimonides permitted the *pilegesh* only to a king, effectively undercutting its legality.[17] Thereafter, most Jewish authorities rejected the *pilegesh* relationship.[18] As a contemporary Reform responsum concludes, "We cannot validate this form of marriage, as it violates our ideals of marriage and the laws of the states or provinces in which we live. It is contrary to the general development of Jewish law in the last eight hundred years."[19]

Even discounting the viability of the *pilegesh* precedent, some contemporary authorities validate sexual activity within an ongoing, loving relationship. More than three decades ago, Rabbi Eugene B. Borowitz, a leading Reform commentator, wrote, "For people who truly love, to express that love in sexual relations cannot be called unethical. ...[A]ny love-making which is true to the love between [them], even if [they] are unmarried, is right."[20] The 1998 Report of the Ad Hoc Committee on Human Sexuality to the Convention of the Central Conference of American Rabbis stated, "In a Reform Jewish context, a relationship may attain a measure of [holiness] when both partners voluntarily set themselves apart exclusively for each other, thereby finding unique emotional, sexual and spiritual intimacy."[21] According to a 1993 Statement by the Reconstructionist Movement, "Jewish tradition speaks of a loving partnership as including the divine presence, and affirms sexual intimacy as an integral aspect of fully intimate relationships. We believe that sexual intimacy between loving partners should be viewed as holy."[22]

Conservative authorities go a considerable distance toward accepting the views of their Reform and Reconstructionist counterparts. Rabbi Elliot N. Dorff, on behalf of The Rabbinical Assembly's Commission on Human Sexuality, designed to educate on the rules of Jewish law as well as the concepts and values underlying these rules, states that committed, loving relationships between adults "can embody a measure of morality, although not the full portion available in marriage."[23] Note the use of the word "morality," not the term "holiness," used by the Reform and Reconstructionist Movements. Dorff asserts, however, that it is difficult to live by the Jewish values relevant to sex in an unmarried relationship.

Although the cohabitating couple is devoted to each other, they often resist taking the next step -- a formal public, lifetime commitment, pledging themselves permanently and exclusively to each other. Getting married is especially problematic for children of divorce. Some experts maintain that the strongest impact of divorce occurs when children become adults, negatively influencing their search for love, intimacy, and a permanent union. These adult children of divorce are considerably less confident than their peers, fearful they will repeat their parents' mistakes. They find commitment elusive; they often migrate from one unsuccessful relationship to another.[24]

Living together is not the same as being engaged or married. Today, we see more clearly the benefits of marriage. It is good for adults (and their children). Married couples do better in terms of physical and mental health, longevity, sex lives, and finances.[25]

Cohabitating couples are less securely anchored. They suffer from weaker commitments to each other. They are more likely to split up than a married couple. They are less likely to remain sexually faithful and exclusive. They are less willing to support each other and endure a rocky period, emotionally or financially, such as illness or unemployment.

As far as more mundane matters, they may not have a joint checking account or otherwise pool their financial resources. They may spend their leisure time differently. Without a public expression of vows, there may be problems of miscommunication of intentions and expectations as well as an inability or an unwillingness to assume an enduring responsibility for each other and for having and educating children. Without a lifetime commitment, they may be less willing to invest in and devote themselves to the relationship -- emotionally, financially, and spirituality.

For Spiritual Judaism, sex involves love, responsibility, and the Divine Presence. There is no a hard-and-fast rule -- one size fits all -- for when sexuality and spirituality come together and the Divine Presence abides with a couple, who achieve the wholeness of life, the uniting of body, mind, and spirit, so important to Spiritual Judaism. For some, it occurs in an ongoing, loving relationship.

However, do most of those in a cohabitating relationship feel that the Divine Presence has entered into their sexual union? Do they believe that their relationship is holy and sacred, somehow connected with the Eternal's Presence? Are they growing and developing their spiritual essence, experiencing God in their lives and in their relationship? From my observation of many in such a relationship, the answer is "No." Each likely does not know the soul of the other. There is not a oneness of their two souls. The physiological sexual act is probably neither linked to higher, more spiritual aspects nor does it operate transcendentally. The absence of permanence in their relationship, no matter how loving, likely indicates that the sexual and emotional aspects lack a spiritual dimension.

Even prior to an engagement, a need exists for some type of a spiritual celebration for cohabitants entering into sexual relations. Spiritual Judaism urges letting them affirm their sexual link through a ceremony, marked by temporary provisional vows, evaluated and renewed annually. A heterosexual couple could recite simple vows before a close circle of family and friends, confirming the relationship, providing it with seriousness and sacredness, but not permanency.[26] If their sexual union produces a child, then, however, they ought to marry.

Post-Engagement Sexual Intercourse

What Spiritual Judaism hopes for as a Jewish sex ethic, for most couples, is post-engagement sexual intimacy. An engagement represents a heightened commitment, although, of course, far less binding than a wedding ceremony with its public expression of vows. An engagement, together with the setting of a wedding date, represents the intent of a man and a woman to foster a permanent relation of love. It represents a commitment, an intention to marry, that a couple is ready to undertake. Although an engagement may be broken more easily than a marriage, its breakup, for most, a painful mini-divorce, is only somewhat easier than a marital dissolution.

Sexual intimacy enables engaged couples to see in advance of marriage that they are well mated for each other. Establishing their sexual compatibility reduces the likelihood of divorce. Readily available contraceptives reduce, but do not eliminate, the risk of pregnancy. It is also realistic in view of the many long engagements today that delay marriage for one year or more.

Until modern times, engagement was the major means by which the Jewish community met the problem of pre-marital sexual relations. In the Jewish experience over the centuries, engaged persons, as a practical matter, had intercourse before marriage.[27] Exclusivity was paramount. An engaged woman was "set aside." She was prohibited from having sexual relations with any other male except her husband-to-be with the full force of the traditional ban against adultery (as discussed in Chapter 6).

In ancient times, sexual activity occurred between a man and a woman who were betrothed to each other. Betrothal, a sort of semi-marriage, took place one year before an actual marriage and established a legal relationship between bride and groom. At that point, sexual intercourse between those who were betrothed was unobjectionable among the ancients in Judea, the southern of the two Jewish kingdoms in the Biblical period, but not in the Galilee, the northern kingdom (Babylonian Talmud Ketubah 7b, 12a).

In Judea, substantial freedom in the sexual relations of engaged couples existed.[28] It was the custom for the man to live with the family of a woman he intended to marry during the year between betrothal and marriage (Mishnah Ketubot 1:5). Although the stricter view prohibiting pre-marital sex prevailed, becoming the norm, such relations apparently remained fairly common during the first four centuries of the Common Era (Babylonian Talmud Yevamot 69b).

In the medieval period, sexual intercourse generally occurred between those who were betrothed. The old Judean custom was reintroduced in Romania, spreading through the Byzantine Empire, and extending even to Southern Italy.[29]

Not unexpectedly, the Jewish tradition's prohibition of sexual intimacy between an engaged couple was difficult to enforce. Although carrying no legal penalty, rabbis considered their behavior "immoral," demanding that engaged couples immediately marry. Sexual activity was viewed as bonding both partners. The rabbis did everything in their power to have Jewish communities stop the practice of permitting an engaged couple's sexual activity.

Although the Reform Movement today considers pre-marital chastity "to be our ideal," one responsum states: "Such [pre-marital] relationships were prohibited and discouraged by authorities throughout the ages. Little was done when such relationships took place between two engaged persons, except in puritanical periods. Other sexual relationships between single adults were prohibited, and every effort was made to enforce such prohibitions. These prohibitions were equally strong upon the man and the woman. In times of lower moral standards, authorities were occasionally permissive or simply looked the other way. Generally, the effort to enforce high moral standards succeeded, and the responsa call attention to the failures." With a note of resignation, the Reform rabbis conclude, "In our period of loose standards, it would be appropriate to do everything within our power to encourage higher standards for both men and women."[30]

Given the modern commitment phobia, formally announcing their engagement and setting a wedding date, marks a special event for most couples. Because many young adults are the products of marriages ending in divorce (or they themselves have been divorced), marriage can be a terrifying event. With an engagement, although not reaching the highest state of public commitment, that is, marriage, there is a sense of reverence beyond oneself. Spiritual Judaism suggests that it is much more likely that the sex lives of engaged couples are invested with attributes of holiness and sanctity (although these attributes may occur earlier in an ongoing, loving relationship). For engaged couples, pre-marital sexual intimacy may bring them closer to their spiritual essence. Their consciousness is more likely directed toward God, linking their sexuality and spirituality.

The acceptance of pre-marital sex, specifically, by engaged, heterosexual couples by Judaism will neither undermine marriage as an institution nor family life. In our self-righteous judgmentalism we should not condemn others who experience sexuality and spirituality at a different point than that advocated by the Jewish Tradition. We need not allow Judaism's single-minded emphasis on heterosexual marriage, family, and children thwart the desire of singles, childless couples, or homosexuals to create meaningful lives for themselves, in terms of sex and procreation.

Let us turn and examine homosexuality in the Jewish tradition, remaining open to the evolution of family units, welcoming a diversity in lifestyles and family arrangements.

Notes

1. I have drawn on Elliot N. Dorff, *"This Is My Beloved, This Is My Friend": A Rabbinic Letter on Intimate Relations* (New York: The Rabbinical Assembly, 1996), 13-16; David M. Feldman, *Birth Control in Jewish Law: Marital Relations, Contraception, and Abortion As Set Forth in the Classic Texts of Jewish Law* (Northvale, NJ: Jason Aronson, 1998), 27-36, 46-53; Robert Gordis, *Love & Sex: A Modern Jewish Perspective* (New York: Farrar Strauss Giroux, 1978), 99-100.
2. Hyam Maccoby, "*Halakhah* and Sex Ethics," in *Dynamic Jewish Law: Progressive Halakhah: Essence and Application*, ed. Walter Jacob and Moshe Zemer (Pittsburgh: Rodef Shalom, 1991), 137-142, interprets this story as an application of the principle against blackmail, not ultra-puritanism.
3. Louis M. Epstein, *Sex Laws and Customs in Judaism* (New York: Bloch, 1948), 31-64.
4. *Id.*, 75-91.
5. *Id.*, 104-123. These rules are summarized by Rabbi Shlomo Ganzfried in *Kitzur Shulchon Oruch: The Classic Guide to the Everyday Observance of Jewish Law*, Volume II, trans. Rabbi Eliyahu Touger (New York: Moznaim, 1991), Chapter 152: The Prohibition Against Yichud (Being Alone with Other Women), 623-626.
6. Maurice Lamm, *The Jewish Way in Love and Marriage* (San Francisco: Harper & Row, 1980), 26, 31-32; Elliot N. Dorff, *Matters of Life and Death: A Jewish Approach to Modern Medical Ethics* (Philadelphia: Jewish Publication Society, 1998), 136.
7. Dorff, *"This is My Beloved,"* 30.
8. Central Conference of American Rabbis (CCAR), Responsum 5756.10 "Long-Term Non-Marital Relationships," available from ccarnet.org.
9. Centers for Disease and Control Prevention (CDC), *CDC Surveillance Summaries*, Youth Risk Behavior Surveillance - United States, 1999 (June 9, 2000) MMWR 2000: 49 (No. 55-5), 75, Table 30. Percentage of high school students who engaged in sexual behaviors, by sex, race/ethnicity, and grade,

United States, Youth Risk Behavior Survey, 1999; CDC, *Morbidity and Mortality Weekly Report*, "Trends in Sexual Risk Behaviors Among High School Students - United States, 1991-1997," 47:36 (September 18, 1998): 749-752, 751, Table 1. Percentage of high school students who reported sexual risk behaviors, by sex, grade, race/ethnicity, and survey year - United States, Youth Risk Behavior Survey, 1991, 1993, 1995, 1997; CDC, *Morbidity and Morality Weekly Report*, "Youth Risk Behavior Surveillance: National College Health Risk Behavior Survey -- United States, 1995," 46(SS-6) (November 14, 1997): 1-54, Table 15. Percentage of college students who reported engaging in sexual behaviors, by sex, age group, race/ethnicity -- United States, National College Health Risk Behavior Survey, 1995.

10. Eugene B. Borowitz, *Choosing A Sex Ethic: A Jewish Inquiry* (New York: Schocken, 1969), 35-49.

11. *The Code of Maimonides*: Book Four: The Book of Women, ed. Isaac Klein (New Haven: Yale University Press, 1972), Treatise 1:4, Laws Concerning Marriage, 5-6; Fred Rosner, M.D., *Sex Ethics in the Writings of Moses Maimonides* (New York: Bloch, 1974), 111.

12. A.E. Waite, *The Holy Kabbalah: A Study of the Secret Tradition in Israel as unfolded by Sons of the Doctrine for the Benefit and Consolation of the Elect dispersed through the Lands and Ages of the Greater Exile* (New Hyde Park, NY: University Books, 1960), 377-402; Borowitz, *Choosing a Sex Ethic*, fn. 72, 162-163.

13. CCAR Responsum No. 154. "Jewish Attitude Toward Sexual Relations Between Consenting Adults (1979)," in *American Reform Responsa: Collected Responsa of the Central Conference of American Rabbis 1889-1983*, ed. Walter Jacobs (New York: CCAR, 1983), 483.

14. *U.S. Census Bureau, Statistical Abstract of the United States: 2000*, 120th edition (Washington: Government Printing Office, 2000), 52, Table No. 57. Unmarried Couples by Selected Characteristics: 1980 to 1999; Jennifer Steinhauer, "Fewer Take Journey to the State of Wedlock," New York Times, January 9, 2000, New York Report, 21, 22; Sara Terry, "The Unexpected Consequence of 'Living Together,'" Christian Science Monitor, April 10, 2000, 1, 5.

15. CDC, Vital and Health Statistics, *Fertility, Family Planning, and Women's Health: New Data From the 1995 National Survey of Family Growth*, Series 23, No. 19 (Washington: Government Printing Office, 1997), 5-6; Pamela J. Smock, "Cohabitation in the United States: An Appraisal of Research Themes, Findings, and Implications," Annual Review of Sociology, vol. 26 (2000): 1-20, at 3; David Popenoe and Barbara Dafoe Whitehead, *The State of Our Unions 2000: The Social Health of Marriage in America* (New Brunswick, NJ: The National Marriage Project, Rutgers University, 2000), 26, 28.

16. Louis M. Epstein, "The Institution of Concubinage Among The Jews," *Proceedings of the American Academy for Jewish Research*, vol. 6 (Philadelphia:

Jewish Publication Society, 1935), 153-188.
17. *The Code of Maimonides*, Book Fourteen, Book of Judges, ed. Abraham M. Hershman (New Haven: Yale University Press, 1949), Treatise 5:4, Kings and Wars, 215; Epstein, "The Institution of Concubinage," 182.
18. Borowitz, *Choosing a Sex Ethic*, 43-50. One authority Rabbi Jacob Emden (1697-1776) argued, in a significant, liberal, but minority, responsum for the revival of *pilegesh* to increase the Jewish population and to keep Jews from even greater sexual immorality. He concluded that it was better for people to engage in sexuality with some degree of religious sanction than to engage in a totally prohibited activity. Gershon Winkler, *Sacred Sex: The Sanctity of Sex in Jewish Law and Lore* (Northvale, NJ: Jason Aronson, 1998), 86-98, 105-142.
19. CCAR Responsum No. 133. "Concubinage as an Alternative to Marriage (1979)" in *American Reform Responsa*, 410.
20. Borowitz, *Choosing a Sex Ethic*, 109.
21. CCAR, Ad Hoc Committee on Human Sexuality, "Report to the CCAR Convention," June, 1998, available from ccarnet.org.
22. The Report of the Reconstructionist Commission on Homosexuality, *Homosexuality and Judaism: The Reconstructionist Position*, rev. ed. (Wyncote, PA: Federation of Reconstructionist Congregations and Havurot and Reconstructionist Rabbinical Association, 1993), 37.
23. Dorff, "*This Is My Beloved*," 35.
24. Judith S. Wallerstein, Julia M. Lewis, Sandra Blakeslee, *The Unexpected Legacy of Divorce: A 25 Year Landmark Study* (New York: Hyperion, 2000).
25. The benefits of marriage are detailed by Linda J. Waite and Maggie Gallagher in *The Case for Marriage: Why Married People Are Happier, Healthier, and Better Off Financially* (New York: Doubleday, 2000).
26. Rosemary Radford Ruether, *Christianity and the Making of the Modern Family* (Boston: Beacon, 2000), 214-215. For practices in other countries see Suzanne Daley, "French Couples Take Plunge Short of Marriage," *New York Times*, April 18, 2000, A1 (French civil solidarity pact) and Elaine Sciolino, "Love Finds a Way in Iran: 'Temporary Marriage,'" *New York Times*, October 4, 2000, A3.
27. Borowitz, *Choosing a Sex Ethic*, 96.
28. Gordis, *Love & Sex*, 166.
29. Epstein, *Sex Laws*, 127-128.
30. CCAR Responsum No. 154. "Jewish Attitude Toward Sexual Relations Between Consenting Adults (1979)," in *American Reform Responsa*, 483.

Chapter 4

Homosexuality and the Jewish Tradition

The heterosexual nuclear family is not the only way that two individuals may meet their sexual, emotional, and economic needs. In twenty-first century America, Jews in many types of families, including gay and lesbian couples, can fulfill the traditional roles discussed in Chapter 3, including serving as the primary unit of intimacy, offering stability, emotional and financial support, and facilitating the transmission of the Jewish tradition.

This chapter focuses on homosexual relations, unanimously condemned by traditionalists. After considering the reasons behind their disapprobation, we turn to more liberal approaches, especially the position of the Reform and Reconstructionist Movements, during the past decade. The reasons for the increasing acceptance of homosexuals and same-sex unions are considered, focusing on modern evidence of causation, choice, and treatment of homosexuality. Spiritual Judaism supports allowing homosexual couples to create the families they choose, unencumbered by the conventional notions of heterosexual marriage.

The Traditional Text-Oriented, Approach to Homosexuality

The Jewish tradition distinguishes male and female homosexuality. Although both are prohibited, both are not equally punishable. Male homosexuality is regarded as an abomination and is categorically prohibited.[1] Female homosexuality is merely prohibited by implication.

Male Homosexuality

For the text-oriented traditionalist, the answer is easy, there is a clear biblical prohibition against gay sex. As Rabbi Moshe Feinstein, one of the twentieth century's leading halakhists, put it, homosexuality has no justification whatsoever, "it is an act of defiance, an abomination par excellence." According to Rabbi Feinstein, "the homosexual act is devoid of any natural sexuality in any form, and does not derive from the divinely implanted sexual urge. It is instead a fundamental expression of rebellion against the biblical norm, almost a deliberate flouting that is carried out *because* the act is forbidden...."[2] Modern Orthodox Rabbi J. David Bleich concluded, "While Judaism regards the homosexual act with repugnance, it has the greatest sympathy for the homosexual as a person.... [but] [s]ociety has a definite obligation not to bestow a seal of legitimacy on homosexual activity."[3]

As discussed in Chapter 2, traditionalists view the Five Books of Moses as the enduring, paramount word of Divine authority. In forbidding male homosexual acts, regarded as a grave sin, the Torah states: "No man is to have sexual relations with another man; God hates that.... They have done an abomination." Capital punishment is ordained for transgressors. We are told: "They both shall be put to death. They are responsible for their own death" (Leviticus 18:22 and 20:13).

The authors of the Torah found any deviation from the heterosexual mode of sexual activity deeply offensive. Simply put, they despised gays, regarding their behavior as abhorrent, something disgusting. Although the Book of Leviticus offers a blanket condemnation of male anal intercourse, it is unclear whether the prohibition relates to discrete gay sex acts as opposed to a committed relationship. Also, is a sexual act an abomination only if it is used as a means of abuse or domination or as the defiant flouting of tradition as Rabbi Feinstein said? Today, traditionalists continue to assert the binding nature of the biblical condemnation of male homosexuality.

The Hebrew Bible also contains various narratives dealing with male homosexuality. It is one of the wicked deeds of the Sodomites. In Genesis 19:5, the men of Sodom demanded to "know," that is, to be intimate with, the two men (actually angels sent by God) visiting Lot, the nephew of the legendary patriarch Abraham and an outsider residing in the town of Sodom. (It is from this account that we derive the term "sodomy.") The townsmen besieged Lot's house and demanded that he turn the two men over to them for nonconsensual sexual intercourse. Instead, Lot offered his two daughters in the place of the male guests, but the townsmen refused them. It became clear that Sodom was beyond hope and we are told that God decreed its destruction (Genesis 19:24-25). In this legend, it is unclear, however, whether God punished Sodom for the attempted gay intimacy or more for its general wickedness and "inhospitality" to strangers.

A somewhat similar experience is recounted in the Book of Judges (Judges 19:22). This time it involved a group from the tribe of Benjamin who sought to commit a homosexual rape.

In a number of other places in the Hebrew Bible (1 Kings 14:24, 15:12) sodomy is encountered in accounts of such practices in the days of the kings of Judea, particularly Rehoboam who likely acted under the influence of his Ammonite mother. It was only finally erased by King Josiah, around 620 B.C.E., who eradicated the role of ritual male homosexual prostitutes (2 Kings 23:7).

In addition to their condemnation of gay sex acts, including the imposition of the ultimate sanction, death by stoning (Babylonian Talmud Sanhedrin 54a-b), the rabbis further warned against writing a marriage contract for male partners (Babylonian Talmud Hullin 92a-b). We are told that Rabbi Huna said in the name of Rabbi Joseph that the generation of Noah was condemned to eradication by the flood because they had sunk so low morally. They wrote male homosexual marriage contracts, publicly legitimizing and making permanent homosexual unions (Leviticus Rabbah 23:9 on Leviticus 18:3, Genesis Rabbah Parashah 26:5 on Genesis 6:2).

In light of the specified prohibition and the mandated punishment, rabbinic sources imply that male homosexuality was neither widespread among ancient Jews nor was it hidden or suppressed. Jerusalem Talmud 4:11 teaches that Rabbi Judah forbade two bachelors from sleeping under the same blanket for fear it would lead to homosexual temptations -- their homosexual tendencies might come to the fore. The sages, however, permitted two bachelors to sleep under the same blanket because homosexuality was so rare among the Jews that preventive legislation was considered unnecessary (Babylonian Talmud Kiddushin 82a), although the Jerusalem Talmud (Sanhedrin 6:3) records male homosexual activity in a Jewish community, accidentally encountered by R. Judah ben Pazzi.

Subsequently, in his Code, Maimonides outlawed male homosexuality.[4] Regarding it as a severe sexual transgression, he repeated the rabbinic punishment, death by stoning for adult gays, both the active and the passive partner.

Female Homosexuality

Lesbianism in the Jewish tradition follows a more circuitous route than the explicit condemnation of male homosexuality. Female homosexuality was not mentioned in the Torah probably because it does not involve the emission of semen or genital intercourse.

In the rabbinic literature, we read about Rabbi Huna opining that a lesbian could not marry a priest, who functioned during the time of the Second Temple in Jerusalem. He reasoned that lesbianism was the equivalent of prostitution (Babylonian Talmud Shabbat 65a). The other rabbis vetoed his stance, viewing lesbianism as an unseemly act of sexual licentiousness (Babylonian Talmud Yevamot 76a). Lesbian relations were viewed as simple lewdness, not prostitution, as no intercourse was involved. Although permitting a lesbian to marry a priest, sages prohibited the act of female homosexuality, viewing it as rebellious behavior and a minor infraction subjecting lesbians to corporal, not capital, punishment for violating the standards of sexual purity.

Despite their somewhat more liberal approach to lesbianism, in contrast to male homosexuality, the Babylonian Talmud (Shabbat 65a) tells us that Samuel's father did not permit his daughters to sleep together, presumably to avoid any sexual intimacy between them.

Maimonides prohibited lesbianism.[5] He held that a Jewish court could order women who engaged in lesbian acts to be whipped.

Reasons Underpinning the Traditional Jewish Approach

Three major reasons exist for the Jewish tradition's prohibition of homosexuality. First, heterosexuality is viewed as maintaining Jewish distinctiveness. Second, homosexuality is seen as undermining marital life and family arrangements, frustrating reproduction, and is unnatural. Third, male homosexuality also violates the prohibition of wasting seed. This section develops and critiques each of these assertions.

First, sages sought to maintain Jewish distinctiveness. One way involved refraining from the sexual practices of the surrounding peoples.

Sexual abomination was viewed, according to the Book of Leviticus (18:24-30), as polluting the land and resulting in the exiling of its inhabitants. Thus, Jews were advised to avoid the practices of neighboring societies.

In arriving at their negative opinion of lesbianism, the rabbis looked at the Torah's prohibition against emulating the practices of ancient Egypt and Canaan (Leviticus 18:3), one of which they alleged, was marriage between two women. In rejecting the lesbian relationships found in these nations, the practices forbidden to Jews may perhaps be seen as those associated with cultic rites and idolatrous worship, not sexual relationships as we know them today.

Second, as we have seen in Chapter 3, the Jewish tradition strives to preserve the one specific type of family structure. This aim has longstanding roots. The authors of the Hebrew Bible and the rabbis saw the heterosexual family, consisting of a husband, a wife, and their children, as the ideal (Genesis 2:18, 24; Genesis Rabbah Parashah 17:1-3, 18:5). Conversely, Jewish sages viewed a homosexual lifestyle as undermining the sanctity of marriage, devaluing the family as a basic social institution, and negating sexuality leading to procreation.

In view of the ancient pattern of teenage marriages, perhaps the rabbis were concerned about the possible breakdown of marital bonds. Medieval Jewish commentators feared a married man with latent homosexual tendencies might, subsequent to his adolescent marriage, later realized his genuine orientation, and, in spite of having created a family unit, abandon his wife and children to pursue a homosexual relationship. If this happened, it could destroy the fabric of heterosexual family life.[6] These fears hold far less weight in view of the contemporary trend to later marriages between mature individuals.

Today, many of us realize that the acceptance of homosexual relationships does not produce moral evil or threaten the family. Opinions have changed on the basis of knowing people who are openly gay or lesbian. Our newfound familiarity with homosexuals shows us that they are decent, loving individuals who do not fit longstanding, negative stereotypes of perverts living a licentious life of moral abomination. Homosexual behavior need not be promiscuous.

We now see more clearly and can accept the notion of an ongoing, loving relationship between members of the same gender. Many homosexuals are involved in constructing family units. They are forming enduring relationships based on love, support and responsibility, emotional intimacy and mutual faithfulness. This type of homosexual couple, monogamous, respectful and supportive of each other, promotes family stability. These exclusive, mutually committed homosexual unions do not undermine heterosexual marriage or the institution of the family.

We need to remember that the vast majority of humanity will continue to be heterosexual. Validating homosexual relationships will not undercut the desire of heterosexual couples to assume their responsibility of childbearing and rearing or the family's cohesiveness and its privileged position.

Also, as examined in Chapter 3, a major function of married heterosexual intercourse is procreation and the affirmation of the continuity of biological life and the Jewish people. Although procreation is impossible through homosexual acts, homosexual unions no longer serve as a barrier to having children.[7] As developed in Chapters 6 and 7, because of technological advances, homosexuals can procreate. New reproductive techniques, such as donor insemination and surrogate motherhood, enable gays and lesbians to form families and raise children of their own. Gays can have children through surrogacy arrangements. Male sperm can artificially impregnate a woman who carries and gives birth to a child, raised by the gay father. Lesbians can have children through artificial insemination, through known donors or anonymous sperm banks. Male sperm from a gay can artificially impregnate a lesbian who bears and raises the child. In addition, homosexual couples can form families with children by adopting children or becoming foster parents.

Homosexuals can rear their children in families. Gay and lesbian families are families -- Jewish families -- providing the love and discipline we hope for. A two-parent gay or lesbian family may provide more benefit than a one-parent household. A homosexual couple's long-term commitment to their relationship and their child rearing is far more important for parenting than their sexual orientation.[8]

These nontraditional family units do not offer impediments to raising "well-balanced" children. No evidence exists that children growing up with gay or lesbian parents, are, overall, less well adjusted, psychologically or socially, than children growing up with heterosexual parents. There are no significant differences in emotional security or interpersonal skills of children raised by heterosexual or homosexual couples.[9] Furthermore, children raised in homosexual households grow up to be heterosexual at substantially the same rates as children raised in heterosexual homes.[10]

Even if homosexuals do not produce offspring or rear children, they can foster the continuity of the Jewish community in other ways, for instance, by being involved in the rearing of their nieces and nephews, or serving as teachers, counselors, or youth advisors. More generally, they can provide the fruits of their special gifts to the whole society. Procreation is not the be all and end all of our earthly existence. Jewish sages remind us that sexual intercourse between husband and wife, even without procreation, is good in itself (Babylonian Talmud Pesahim 72b).

Third, the ancients evidenced concern about the spilling of male seed. Semen can be spilled through masturbation, gay sexual acts, and heterosexual acts resulting in non-vaginal ejaculation.

Although the Torah does not prohibit masturbation, the Book of Genesis tells the story of Onan, the son of Judah, who spilled his seed on the ground rather than impregnate his brother's childless widow, which it was his obligation to do under the institution of levirate marriage (Deuteronomy 25:5-10), thereby depriving his brother's line of progeny. Displeased by Onan's act, perhaps because Onan's act lessened the Jewish population or because of Onan's disobedience, God, we are told, killed him (Genesis 38:7-10).

Although Onan's act may be seen as the abuse of a woman, the Babylonian Talmud (Yevamot 34b) refers to Onan's spilling of seed as unnatural. A sense of the unclean or because it was seen as lessening a man's need for intimacy with a woman seems to have motivated the prohibition against masturbation by the school of Rabbi Ishmael (Babylonian Talmud Niddah 13b). One who brought forth semen in vain was said to deserve death; it was considered a sin punishable by God (Babylonian Talmud Niddah 13a).

Both the Code of Maimonides[11] and Karo's *Code of Jewish Law* condemn and prohibit masturbation. The *Code of Jewish Law* states: "It is forbidden to discharge semen in vain. The sin is more severe than any of the other prohibitions of the Torah. Those who masturbate and thus discharge semen in vain not only violate a severe prohibition, but they are under a ban. Concerning them it is said (Isaiah 1:15), 'Your hands are full of blood.' It is equivalent to murder."[12]

The principal book of the Jewish mystical tradition, the *Zohar* (II 219b) crystallized the medieval sense of horror at masturbation by threatening punishment on the day of reckoning. Through this sin, we are told, "one pollutes himself more so than through any other sin in this world or the next." In the afterlife he is "thrust down lower than others" in the World to Come, never having a chance to ascend. Thus, ancient and medieval Jews saw masturbation as the supreme disrespect for the gift of being allowed to participate in the Divine act of creation.

Today, we do not associate the creation of human life with the moment of ejaculation. And furthermore, know that the fertilization process involves both an egg and a sperm.

Liberal authorities do not regard masturbation as harmful or sinful. Because it hurts no one and may help teenagers abstain from intercourse, allowing them to mature and engage in sex with sanctity and responsibility, they even permit it, at least in moderation.[13] The ancient prohibition against spilling seed, therefore, should not control the Jewish position on male homosexuality. Furthermore, female homosexuality does not entail the spilling of seed.

Finally, accepting (and even sanctifying) loving homosexual relationships will not legitimize incestuous or adulterous relationships. Although R. Akiba (Babylonian Talmud Sanhedrin 58a) derives the prohibition of incest (Leviticus 18:6-18), adultery (Leviticus 18:20), and homosexuality (Leviticus 18:22) from different parts of the same biblical verse (Genesis 2:24), there is no slippery slope. These comprise three separate problems. We can accept homosexuals and homosexual unions without approving of incest or adultery, both typically involving an abuse of a position of power or a relationship of trust.

Liberal Jewish Approaches to Homosexuality

The more progressive branches of Judaism, Reform and Reconstructionist, accept homosexuals, thereby breaking with traditionalists. As is typical of many mainstream religions in the United States, deep divisions exist in the Conservative Movement.

The Conservative Movement's approach to homosexuality, although extending a measure of tolerance and support for homosexuals, remains rooted in traditional, textual Jewish sources. In 1990, The Rabbinical Assembly, the professional organization for Conservative rabbis, passed a resolution supporting civil rights for homosexuals, welcoming gay and lesbian Jews into Conservative synagogues, deploring anti-gay and lesbian violence, and calling on the movement to increase its "awareness, understanding and concern" for Jews who are gay or lesbian.[14]

The 1990 resolution was followed in 1992 by the Conservative

Movement's Committee on Jewish Law and Standards adopting a series of responsa taking two different approaches to homosexuality. In brief, one group declared that homosexuality should no longer be considered as abomination, leaving unresolved how to reconcile an openness to gay and lesbian Jews with the commitment to Jewish family values.[15] According to another group of Conservative rabbis, homosexual acts remain forbidden and homosexuals should remain celibate.[16] Reflecting its adherence to historical Jewish sources, this second group and the Conservative Movement generally does not, at present, want to take any act that could "reasonably be understood to imply halakhic coequality, validation, or acceptability of a homosexual lifestyle."[17]

Given this division of opinion, in 1992 the Committee on Jewish Law and Standards affirmed three basic policies.[18] First, Conservative rabbis may not perform commitment ceremonies for gays or lesbians, although some rabbis do and are allowed to remain in The Rabbinical Assembly. Second, the Conservative Movement is unwilling to accept openly gay or lesbian rabbis; avowed homosexuals cannot be admitted to Conservative rabbinical schools or The Rabbinical Assembly. Acceptance of openly homosexual rabbis and their lifestyle would imply that homosexuality is halakhically acceptable.[19] Third, although affirming that homosexuals are welcome in Conservative synagogues, the Committee charged the rabbi of each Conservative congregation, in consultation with its lay leaders, to formulate policies regarding the eligibility of homosexuals for honors within worship services and for lay leadership positions. Each synagogue's rabbi is given the authority to decide whether open homosexuals may function as teachers or youth leaders in the congregation and its school.

The Reform Movement's viewpoint on homosexuality has undergone a profound transformation over the past 30 years, culminating in seeing the potential for sacredness (*kiddushin* in Hebrew) in same-gender sexuality. In a 1973 responsum, Reform Rabbi Solomon B. Freehof considered homosexuality a "grave sin." Viewing the homosexual mode of life as "a contravention of all that is respected in Jewish life," he disallowed any possibility of sacredness in the relationship of two gays or two lesbians. However, he concluded it was wrong to isolate homosexuals into a separate congregation or exclude them from normal congregational activities.[20]

In answering a question of whether congregations could engage a known homosexual as a religious school teacher or an executive secretary, a 1981 responsum evidenced a negative attitude. This responsum concluded: "[O]vert homosexual behavior which is considered objectionable by the community disqualifies the person involved from leadership positions in the Jewish community. We reject this type of individual as a role model within that Jewish community. We cannot recommend such an individual as a role model nor should he/she be placed in a position of leadership or guidance for children of any age."[21]

Subsequently, a 1986 responsum evidenced a somewhat greater acceptance of homosexuals and their children, however, it did so in a backhanded way. This responsum concluded that a lesbian couple could participate fully as parents in a service celebrating the bar mitzvah of the child, adopted by one and raised by both, who was formally converted to Judaism. If they are not "flagrant about their relationship and [do not] make an issue of it," then their "relationship is irrelevant." They could participate in the service as any parents, indicating "to both the congregation and this household that we recognize the love and care given to the child and do not focus on or recognize the lesbian relationship."[22]

To make a long and convoluted story short, over the past twenty-five years the Reform Movement has gradually embraced and signified its approval of homosexuality and homosexual unions. Beginning in 1977, the Central Conference of American Rabbis (CCAR), the professional organization for Reform rabbis, adopted a resolution calling for legislation decriminalizing homosexual acts between consenting adults and an end to discrimination against homosexuals as persons.[23]

A marked shift occurred in the decade of the 1990s. In 1990, the CCAR endorsed a report of its Ad Hoc Committee on Homosexuality and the Rabbinate, urging that all rabbis, regardless of sexual orientation, be accorded an opportunity to fulfill their chosen sacred vocation and that all rabbinic graduates of the Hebrew Union College-Jewish Institute of Religion be admitted to CCAR membership on application. Endorsing the view that "all Jews are religiously equal regardless of their sexual orientation," this report called on "rabbis and congregations to respect and integrate fully all Jews into the life of the community regardless of sexual orientation." However, a majority of the committee reaffirmed unequivocally the centrality of heterosexual, monogamous, procreative marriage as the ideal human relationship and its special status as *kiddushin* (a traditional Jewish marriage). A minority dissented, affirming the possibility of covenantal fulfillment in homosexual and heterosexual relationships with the relationship, not the gender, determining its Jewish value.[24]

Later in the decade of the nineties, by a resolution adopted in 1996, the CCAR indicated its support of the right of homosexual couples to share in the rights of civil marriage. The CCAR voiced its opposition to governmental efforts to ban homosexual marriages.[25]

Subsequently, two CCAR committees addressed the issue of Reform rabbis officiating at same-gender ceremonies. The CCAR Committee on Responsa, by a 7 to 2 vote (acknowledging a lack of consensus), concluded that homosexual relationships, even if exclusive and committed, do not qualify as a Jewish marriage. Likewise, a majority of the committee did not interpret its responsibility as rabbis to warrant officiation at weddings or commitment ceremonies for gay and lesbian couples because *kiddushin* is "an institution whose legal essence excludes homosexual relations. The performance of a ceremony that resembles but is not [*kiddushin*] does not qualify as a Jewish marriage, even if the couple regards it as such."[26]

An interim 1998 Report of the Ad Hoc Committee on Human Sexuality to the CCAR convention concluded that *kiddushah* (holiness) may be present in a committed, same gender relationship between two Jews. These relationships can, the committee noted, serve as the foundation of stable Jewish families, thereby strengthening the Jewish community. Furthermore, the committee believed that the relationship of a same-gender Jewish couple is worthy of affirmation by an appropriate Jewish ritual, but each rabbi is free to decide about officiating according to his or her informed rabbinic conscience.[27]

The acceptance and recognition of same sex unions, prodded by the Women's Rabbinic Network, culminated in 2000. The CCAR adopted a resolution stating that the relationship of a Jewish, same gender couple is "worthy of affirmation through appropriate Jewish ritual," and called for developing of appropriate educational and liturgical resources for ceremonies. Recognizing that diversity of opinion among Reform rabbis on this issue, the resolution indicated support for the decision both of those rabbis who choose and those who choose not to officiate at rituals for the unions of same-gender couples.[28]

Preceding the latest action of the CCAR by some seven years, in 1993 the Reconstructionist Commission on Homosexuality, comprised of lay and rabbinic representatives of the three arms of the Reconstructionist Movement, namely, the Reconstructionist Rabbinical Association, The Federation of Reconstructionist Congregations and Havurot, and the

Reconstructionist Rabbinical College, offered these guidelines:

> We affirm the importance of loving, caring, intimate relationships as a primary source of companionship and comfort. Jewish tradition accords committed relationships which are invested by Jewish ritual with the value of *kedushah*, holiness. We affirm the qualities of mutual respect, trust, care, and love in committed relationships regardless of sexual orientation. Same-gender partnerships have the same potential for embodying these qualities as do heterosexual marriages. As we support the long-term commitment of heterosexual couples and acknowledge the *kedushah* of their marriages, so do we support long-term partnerships between gays or lesbians and affirm that *kedushah* resides in committed relationships between same-gender Jewish couples.
>
> ...
>
> We affirm the existing position of the Reconstructionist Movement that accepts gays and lesbian Jews to the rabbinate.... Despite a lack of public affirmation, gay and lesbian rabbis have led with dignity and distinction, serving all the needs of Jews from birth to death.[29]

Also in 1993, the Reconstructionist Rabbinical Association in addition to welcoming gay and lesbian Jews, their partners and families as full members of Reconstructionist Movement organizations and affirming a commitment to a full place for gays and lesbians as members of Reconstructionist organizations, adopted these guidelines:

> We recognize that some same-gender couples choose to affirm publicly their commitment to one another through Jewish ritual. To accomplish this, they often seek the help of rabbis. In responding to their requests, we recognize pluralism within our ranks and acknowledge that the issue of rabbinic officiation at same-gender commitment ceremonies requires sensitive judgments best left to individual rabbis. We support those of our members who choose to officiate at same-gender commitment ceremonies. We acknowledge that rabbis who support these guidelines may nevertheless choose not to officiate at same-gender commitment ceremonies. We affirm their right to make individual judgments in this complex area.[30]

What Is Propelling the Acceptance of Homosexual Unions?

In addition our newfound realization that homosexual unions will not undermine heterosexual marriage or family life, modern biological findings, summarized in this section, have propelled a rethinking of homosexuality. Despite the increasing popularity of biological theories of sexual orientation, biology is not sexual desire. Biology gets us only so far. With the origins of sexual orientation remaining shrouded in mystery, we need to return to virtues underpinning Spiritual Judaism. We must be compassionate toward others and recognize that every human being is created in God's image.

The authors of the Hebrew Bible, as did the ancient rabbis, assumed that male homosexual behavior represented an act of free choice. The question is, however, more complex. There are three interrelated issues: causation; choice; and the treatment of homosexuality.

No agreement exists in the modern scientific literature as to the cause of homosexuality. Recent research findings emphasize neurological and genetic factors over which a person has no control as the major elements making him gay. Two studies are noteworthy, one neurological and the other genetic.

In a study published in 1991, a scientist found, based on postmortem examinations of the brains of forty-one cadavers, mainly AIDS victims, that an area of the brain, the anterior hypothalamus, a region governing sexual behavior, is smaller in gay men than in heterosexual men. Although cautiously concluding that "the results do not allow one to decide if the size of [this area of the brain] in an individual is the cause or consequence of that individual's sexual orientation, or if [its] size... and sexual orientation covary under the influence of some third, unidentified variable," the author stated that his study illustrates that "sexual orientation in humans is amenable to study at the biological level...."[31]

Building on the premise that genes may contribute to male homosexuality, research efforts have focused on trying to identify the "gay gene." A 1993 study of forty pairs of gay brothers from families with maternal gay relatives showed that these homosexual men carried an identical segment of the X chromosome (Xq28), suggesting that males who inherit this section of the X chromosome from their mothers are predisposed to be gay.[32] Using DNA samples from pairs of gay brothers and, if available, from their mothers, researchers found that eighty-two percent of these 40 pairs had the same genetic sequence in a particular portion of one region of the X chromosome. This study suggested that the q28 region is the place where this difference in sexual orientation is inscribed. Although the study showed that at least one subtype of male sexual orientation is genetically influenced, it did not identify the "gay gene" or even the particular genetic sequence associated with homosexuality.

Attempts at independent replication of this study's linkage analysis have failed to find evidence of an X-linked gene underlying male homosexuality. In a subsequent study of fifty-two pairs of gay brothers, using methods similar to the 1993 study, another group of researchers "found no evidence of linkage of sexual orientation of Xq28...."[33] The gay brothers in this latter study were no more likely to share the Xq28 markers than would be expected by chance.

Where do these empirical studies of causation leave us? Those looking solely or primarily to biological factors, such as, neurological, genetic, or hormonal, assert that one's sexual orientation is inborn or determined at an early age. Because it is hardwired into one's brain, these biological factors determine, or at the very least, strongly constrain, one's sexual orientation.

At present, the empirical evidence does not support a direct model positing that a particular gene (or genes or other biological factors) influence brain structures underlying sexual orientation, with these neurological structures directing one's sexual attraction developing at an early age. We need to recognize that there are "many intervening pathways between a gene and a behavior and even more between a gene and a pattern that involves both thinking and behaving." For the term "gay gene" to have "a clear meaning, one needs to propose that a particular gene, perhaps through a hormonal mechanism, organizes the brain specifically to support the desire to have sex with people of the same sex. No one has, however, presented evidence in support of such a simple and direct link between genes and sexual orientation."[34]

According to our best current understanding, homosexuality results from a combination of biological, psychological, and social factors, perhaps varying in significance in each of us. We do not have one unified theory explaining what causes differences among humans with respect to sexual orientation. It is multifactorial, with no single factor accounting for its existence or expression. Simply put, orientation is a product of nature and nurture with multiple pathways to sexual orientation. As one expert concludes, "...[It] is likely that there are *multiple origins* to human sexual orientations and sexual desires."[35]

According to the indirect model of biological causation, genes code for, or other biological factors influence, psychological characteristics, affecting temperament and personality, and that these psychological qualities, in various contexts and given certain environmental inputs, shape one's sexual orientation. In this view, biological factors, such as one gene or a set of genes, "might predispose [one] to homosexuality in some environments, to heterosexuality in others, and have no impact on sexual orientation in others."[36]

Psychological and social factors play an important role in the development of sexual orientation. One's various experiences (and the subjective interpretations of these experiences) lead different individuals to same or opposite gender attraction. At least for women, "social factors (such as the effects of sexism and the influence of feminism) play a significant role in shaping sexual desires and sexual orientations."[37] The choice of lesbianism may be as much a feminist statement as a sexual one.

In view of the complex of biological, psychological and environmental factors influencing sexual orientation, with diverse factors playing different roles for various individuals, we need to return to a key question: is homosexuality a matter of choice? Most rational men and women would not choose homosexuality considering the past and present social penalties involved. Because of the discrimination and lingering prejudice homosexuals face, most individuals would not select homosexuality over heterosexuality. As one expert states, "...a person does not choose his or her sexual orientation -- either genetic makeup determines sexual orientation or environmental conditions do. However a person's sexual orientation develops, choice is not involved."[38]

Choices in other areas can have the unintended effect of influencing sexual orientation. Thus, this same expert concludes, "Even though a person cannot choose his or her sexual orientation, a person's choices can affect his or her sexual orientation; we just do not know which choices play a role...."[39] One's indirect choices, perhaps even unconscious choices, may play some role in the development of one's sexual orientation.

Finally, therapeutic interventions aimed at altering sexual orientation generally have not met with success. At present, homosexuality seems irremediable and irreversible, at least for most people. It is rare that homosexuals become heterosexuals even with the best available treatment methods. Attempts to change sexual orientation through various therapies, including psychoanalysis, prayer, electric shock or hormonal injections, "have been dramatic failures,"[40] thereby establishing that sexual orientation is resistant to change.

We can offer a few tentative conclusions regarding the questions of causation, choice, and treatment. Homosexuality is an orientation, a product of complex causal factors rendering it part of one's psychological makeup, probably not a conscious choice on the part of most individuals, especially males. Homosexuality represents a basic psychic orientation, involving the deepest levels of one's personality. Although an individual's choices may impact on one's sexual orientation, it is generally beyond one's control. Furthermore, homosexuality is generally not amenable to any known form of treatment.

Spiritual Judaism and Homosexuality

If we look back to the interpersonal virtues developed in Chapter 2, heterosexuals ought to view with compassion the plight of homosexuals in our society. We need to offer words and acts of kindness towards them, showing acceptance and including them in our circle of neighborly love. We are told: "Love your neighbor as yourself..." (Leviticus 19:18). The phrase refers to our fellow human beings, not just to one's blood relatives or friends.

Spiritual Judaism counsels feeling kinship with one who is marginalized, making friends with and having compassion for the stranger, someone who is different, striving to know his or her heart and to make him or her feel at home with us and others. In seeing a homosexual as a human being with hopes, dreams and aspirations, helps us discover our own humanity.

Yet, this way of pleading the virtue of compassion may not commend itself to everyone. Traditionalists, especially, are loyal to their understanding of the Torah and its abhorrence of homosexuality. Some strive to express their compassion within that context. For Modern Orthodox Rabbi Barry Freundel:

> "...Compassion needs to recognize that homosexuality is fundamentally narcissistic, a focus on the same as opposed to the other....
>
> Compassion, therefore, does not argue for acceptance of homosexuality. Certainly, gay bashing and inappropriate discrimination are wrong and should be condemned as strongly as possible, but it is also wrong to present a homosexual lifestyle as equivalent to a heterosexual lifestyle.
>
> ...Compassion calls for creating an environment that encourages change, punishes gay bashing and understands that change may be very difficult."[41]

Spiritual Judaism argues that despite the literal words of the Torah, we need to recognize that every human is created in the Eternal's image. We are all equal in the Eternal's eyes, even those who are humiliated, despised, ridiculed, mocked, rejected, or condemned by the majority of society. Sexual orientation is irrelevant to a human's worth. God cares for all of us regardless of our sexual orientation. The Divine Presence flows freely through each of our hearts.

Traditionalists would require celibacy of homosexuals,[42] forcing them to live an emotionally restricted life and refrain from sexual expression. This seems "cruel" and life denying.[43] God, Who is good and merciful, does not demand that homosexuals remain celibate, sexually frustrated. Sexuality is a gift from the Eternal and its pleasures should not be denied to anyone acting in the context of a loving, spiritually-filled relationship. Spiritual Judaism does not condemn homosexuals to a life of incompleteness, to denying their emotional and sexual feelings.

All people, regardless of their sexual orientation, deserve respect. Spiritual Judaism suggests letting homosexuals confirm and express their love; let them affirm their sexuality; let us accept all people living in committed, loving, supportive, and respectful relationships.

Just as with heterosexual couples, prior to any expression of sexuality, we need to ask: is the particular homosexual relationship intended to foster an ongoing relationship of love and nurturance in which a spiritual dimension is present? Some may find the connection between sexuality and spirituality in a loving and caring relationship, a meaningful partnership between two humans. For other homosexual couples, just as with heterosexual couples, Spiritual Judaism recommends that some type of public announcement, such as an engagement or a new form of spiritual celebration, marked by the exchange of temporary vows, annually evaluated and renewed (as discussed in Chapter 3), may help link sexuality and spirituality, thereby enabling a couple to connect with the Divine Presence and link together sex and the sacred.

Spiritual Judaism thus supports same-sex marriages. Other states should follow the lead of Vermont in recognizing civil unions for homosexuals.[44] In Vermont, a civil union, paralleling marriage, legally recognizes the lifelong commitments of a man to a man or a woman to a woman. Civil unions are certified by a justice of the peace, a judge, or a member of the clergy. Although, at present, the Federal government will not recognize these civil unions with regard to Social Security benefits and federal taxes, among other items, civil unions grant homosexuals the same rights and protections in Vermont as if they were married, from inheritance rights to property transfers.

Some type of public ceremony, marking the beginning of an exclusive, committed relationship for homosexual couples will help them form more permanent monogamous unions.[45] Realizing that if people want to be faithful to each other, they can do so without legal support; however, some type of legally recognized ceremony may reduce the number of homosexuals living in a promiscuous fashion. For the community at large, it will reinforce our commitment to the family unit by legitimizing stable homosexual relationships. Encouraging enduring homosexual relations will not weaken the fabric of heterosexual family life. Rather, it will enhance the family in all its contemporary configurations. Heightened commitments are good for homosexual and heterosexual couples, offering economic benefits for the family unit and presenting additional barriers to breakups -- the expenses and complications of divorce and its adverse impact on children. For both heterosexuals and homosexuals, the gravity of marriage provides meaningful sanctification and legitimacy for the parties.

Spiritual Judaism urges that we accept a variety of family forms, beyond the traditional heterosexual, nuclear family. Welcoming diversity, these alternative family units, including gay and lesbian couples, grounded in love and responsibility provide a widened choice in the creation and fulfillment of the human connection, fostering mutual support and meeting the need for companionship and emotional intimacy. These new family arrangements help those who partake of them to realize the Divine by enabling them to experience God in their lives and in intimate relations with one another. If Judaism supports them in the right way, they can also contribute to the continuity of the Jewish tradition.

One final point as we consider these new family configurations, especially the children of gay or lesbian unions. Children born of marital heterosexual intercourse do not have a unique claim to being made in the image of God. As discussed in Chapter 2, Divine revelation of the Torah and the rabbinic literature never occurred. God never declared that humanity procreate new life only by heterosexual marriage. Prior to the twentieth century, there was no alternative. But we are now facing new possibilities, discussed in Chapter 6 and 7, for which there is no specific Divine instruction. Therefore, Spiritual Judaism looks to the virtues of love, compassion and responsibility as we seek to foster life and personal fulfillment.

Notes

1. Basil F. Herring, *Jewish Ethics and Halakhah For Our Time: Sources and Commentary: Volume I* (New York: KTAV, 1984), 175-196; J. David Bleich, *Judaism and Healing: Halakhic Perspectives* (New York: KTAV, 1981), 69.
2. Herring, *Jewish Ethics and Halakhah,* 185.
3. Bleich, *Judaism and Healing,* 71-72.
4. *The Code of Maimonides,* Book Five, The Book of Holiness, trans. Louis I. Rabinowitz and Philip Grossman (New Haven: Yale University Press, 1965), Treatise I, Laws Concerning Forbidden Intercourse, Chapter 1:14, 13; Fred Rosner, M.D., *Sex Ethics in the Writings of Moses Maimonides* (New York: Bloch, 1974), 97-98.
5. *The Code of Maimonides,* Book Five, The Book of Holiness, Treatise I, Law Concerning Forbidden Intercourse, Chapter 21:8, 135; Rosner, *Sex Ethics,* 101. For an analysis of the rabbinic and code prohibitions against lesbianism see Rebecca Alpert, *Like Bread on the Seder Plate: Jewish Lesbians and the Transformation of Tradition* (New York: Columbia University Press: 1997), 29-34.
6. John Lewis Eron, "Homosexuality and Judaism" in *Homosexuality and World Religions,* ed. Arlene Swidler (Valley Forge, PA: Trinity Press International, 1993), 113.
7. John Leland, "O.K., You're Gay. So? Where's My Grandchild?" *New York Times,* December 21, 2000, D1.
8. Laura Benkov, *Reinventing the Family: The Emerging Story of Lesbian and Gay Parents* (New York: Crown Publishers, 1994), discusses, among other topics, what homosexual couples tell their children about their relationship and parentage.
9. Fiona L. Tasker and Susan Golombok, *Growing Up In A Lesbian Family: Effects on Child Development* (New York: Guilford Press, 1997), 134-144; Susan Golombok, "New Family Forms: Children Raised in Solo Mother Families, Lesbian Mother Families, and in Families Created by Assisted Reproduction," in *Child Psychology: A Handbook of Contemporary Issues,* ed. Lawrence Balter and Catherine S. Tamis-LeMonda (Philadelphia: Psychology Press, 1999), 439-440; Fiona Tasker and Susan Golombok, "Adults Raised as Children in Lesbian Families," American Journal of Orthopsychiatry 65:2 (April 1995): 203-215; Susan Golombok and Fiona Tasker, "Children in Lesbian and Gay Families: Theories and Evidence," Annual Review of Sex Research 5 (1994): 73-100, at 84-87, 94-95; Julie Schwartz Gottman, "Children of Gay and Lesbian Parents" in *Homosexuality and Family Relations,* ed. Frederick W. Bozell and Marvin B. Sussman (New York: Haworth, 1990); Richard Green et al, "Lesbian Mothers and Their Children: A Comparison With Solo Parents Heterosexual Mothers and Their Children," *Archives of Sexual Behavior* 15:2 (April 1986): 167-184; Susan Golombok et al, "Children in Lesbian and Single Parent Households: Psychosexual and Psychiatric Appraisal," *Journal of Child Psychology and*

Psychiatry 24:4 (October 1983): 551-572. See generally, William N. Eskridge, Jr., *The Case For Same-Sex Marriage: From Sexual Liberty to Civilized Commitment* (New York: The Free Press, 1996), 112-113, fn. 66, 245-246.
10. Tasker and Golombok, *Growing Up in a Lesbian Family*, pp. 102-133; Golombok, "New Family Forms, 440-441; Susan Golombok and Fiona Tasker, "Do Parents Influence the Sexual Orientation of Their Children? Findings From a Longitudinal Study of Lesbian Families," *Developmental Psychology* 32:1 (January 1996): 3-11; Tasker and Golombock, "Adults Raised as Children in Lesbian Families;" Golombok and Tasker, "Children in Lesbian and Gay Families," 80-81. For a review of the literature see Judith Stacey and Timothy J. Biblarz, "(How) Does The Sexual Orientation of Parent Matter," *American Sociological Review* 66:2 (April 2001): 159-183, who note that studies suggest that the experience of having two same gender parents and of growing up in a home where homosexuality is accepted influence children's behavior (specifically, their attitudes toward sexuality and sexual behavior), self-image (how flexibly children interpreted gender roles), and life goals. Other literature reviews include Charlotte J. Patterson, "Children of Lesbian and Gay Parents," *Child Development* 63 (October 5, 1992): 1025-1042.
11. *The Code of Maimonides*, Book Five, Book of Holiness, Treatise I, Laws Concerning Forbidden Intercourse, Chapter 21:18, 137.
12. Rabbi Shlomo Ganzfried, *Kitzur Shulchon Oruch*, Volume II, trans. Rabbi Eliyahu Touger (New York: Moznaim, 1991), 621 (Chapter 151:1).
13. Central Conference of American Rabbis (CCAR) Responsum No. 153. "Masturbation (1979)," in *American Reform Responsa: Collected Responsa of the Central Conference of American Rabbis 1889-1983*, ed. Walter Jacob (New York: CCAR, 1983), 480; Elliot N. Dorff, *Matters of Life and Death: A Jewish Approach to Modern Medical Ethics* (Philadelphia: Jewish Publication Society, 1998), 119-120; The Report of the Reconstructionist Commission on Homosexuality, *Homosexuality and Judaism: The Reconstructionist Position*, rev. ed. (Wyncote, PA: Federation of Reconstructionist Congregations and Havurot and Reconstructionist Rabbinical Association, 1993), 38.
14. *Proceedings of The Rabbinical Assembly's 1990 Convention* (New York: Rabbinical Assembly), 275.
15. Rabbi Elliot Dorff, "Jewish Norms for Sexual Behavior: A Responsum Embodying a Proposal," March 1992.
16. Rabbi Joel Roth, "Homosexuality," November 1991.
17. *Id.*, 118; Rabbi Roth's *teshuvah* was adapted, as was Rabbi Dorff's, as the philosophical and legal rationales for policies on homosexuality adopted by The Committee on Jewish Law and Standards of The Rabbinical Assembly, March 25, 1992.
18. The Committee on Jewish Law and Standards of The Rabbinical Assembly, March 25, 1992.
19. Roth, "Homosexuality," 123.

20. CCAR Responsum No. 13. "Judaism and Homosexuality (1973)," in *American Reform Responsa*, 49-52. CCAR Responsum No. 201 "Homosexual Marriage (1985)," in Walter Jacob, *Contemporary American Reform Responsa* (New York: CCAR, 1987), 298, concludes, "...We cannot accommodate the relationship of two homosexuals as a 'marriage' within the context of Judaism, for none of the elements of...[sanctification] normally associated with marriage can be involved for this relationship."
21. CCAR Responsum No. 14. "Homosexuals in Leadership Positions (1981)," in *American Reform Responsa*, 52-54.
22. CCAR Responsum No. 200. "Lesbians and Their Children (1986)," in *Contemporary American Reform Responsa*, 297.
23. Rights of Homosexuals, Resolution Adapted by the CCAR at the 88th Annual Convention of the CCAR (1977).
24. Report of the Ad Hoc Committee on Homosexuality and the Rabbinate, Resolution Adopted by the 101st Annual Convention of the CCAR (June 1990).
25. On Gay and Lesbian Marriage, Resolution Adopted by the 107th Annual Convention of the CCAR (March 1996).
26. CCAR Responsa Committee, "On Homosexual Marriage," Responsum 5758.6, reprinted in CCAR Journal: A Reform Jewish Quarterly 46:1 (Winter 1998): 5-35, at 26.
27. Report of the CCAR Ad Hoc Committee on Human Sexuality, Report to the CCAR Convention, June, 1998.
28. Resolution on Same Gender Officiation, Resolution adopted at the 111th Annual Convention of the CCAR, March, 2000; Gustav Niebuhr, "Reform Rabbis Back Blessing of Gay Unions," *New York Times*, March 30, 2000, A1; Hanna Rosin, "Reform Rabbis Sanction Gay Unions," *Washington Post*, March 30, 2000, A1.
29. The Report of the Reconstructionist Commission on Homosexuality, *Homosexuality and Judaism*, 37, 39. In 1984, preceding the action of the CCAR by six years, the Reconstructionist Rabbinical College (RRC) established a non-discriminatory admissions policy on the basis of sexual orientation. From its founding in 1975, the Reconstructionist Rabbinical Association (RRA) automatically welcomed all RRC graduates for membership. In 1991, the RRA adopted a formal policy of non-discrimination in membership and prohibited discrimination in the referral of rabbinic resumes for congregational positions. *Id.*, fn 5, 6-7, 39-40.
30. *Id.*, 40-41.
31. Simon LeVay, "A Difference in Hypothalamic Structure Between Heterosexual and Homosexual Men," *Science* 253:5023 (August 30, 1991): 1034-1037, at 1036. Edward Stein, *The Mismeasure of Desire: The Science, Theory, and Ethics of Sexual Orientation* (New York: Oxford University Press, 1999), 196-201, 206, 215, provides a critique of LeVay's study based on the assignment of sexual orientation and a failure to prove causation between size of an area of

the brain and sexual orientation or more generally, that biological factors directly affect sexual orientation. See also William Byne, "The Biological Evidence Challenged," *Scientific American* 270:5 (May 1994): 50-55, at 53-54.
32. Dean Hamer et al, "A Linkage Between DNA Markers on the X Chromosome and Male Sexual Orientation," *Science* 261:5119 (July 16, 1993): 321-329. The linkage was confirmed in a second study of 33 new families with gay brothers. Stella Hu et al, "Linkage Between Sexual Orientation and Chromosome Xq28 in Males But Not in Females," *Nature Genetics* 11:3 (November 1995): 248-256. For a critique of a linkage analysis in the area of genetically complex trait, the subject pool for the study, and the failure to include gay subject's heterosexual brothers, see Stein, *The Mismeasure of Desire*, 219-221. See also William Byne, "The Biological Evidence Challenged," at 54-55 and Terry R. McGuire, "Is Homosexuality Genetic? A Critical Review and Some Suggestions," *Journal of Homosexuality* 28:1/2 (1995): 115-145, at 132-134.
33. George Rice et al, "Male Homosexuality: Absence of Linkage to Microsatellite Markers at Xq28," *Science* 284:5414 (April 23, 1999): 665-667.
34. Udo Schüklenk et al, "The Ethics of Genetic Research On Sexual Orientation," *The Hastings Center Report* 27:4 (July-August 1997): 6-13, at 8. See also Edward Stein et al, "Sexual Orientation," in *Encyclopedia of Applied Ethics*, ed. Ruth Chadwick (San Diego: Academic Press, 1998), 103.
35. Stein, *The Mismeasure of Desire*, 251. See also William Byne and Bruce Parsons, "Human Sexual Orientation: The Biological Theories Reappraised," *Archives of General Psychiatry* 50:3 (March 1993): 228-239, at 236-237; Terry S. Stein, "Deconstructing Sexual Orientation: Understanding the Phenomena of Sexual Orientation," *Journal of Homosexuality* 34:1 (1997): 81-85; Claudia Dreifus, "A Conversation with Anne Fausto-Sterling: Exploring What Makes Us Male or Female," *New York Times*, January 2, 2001, D3.
36. Stein, *The Mismeasure of Desire*, 126-127. See also *id.*, 251-254.
37. *Id.*, 252. See also Celia Kitzinger and Sue Wilkinson, "Transition From Heterosexuality to Lesbianism: The Discursive Production of Lesbian Identities," Developmental Psychology 31:1 (January 1995): 95-104; Carla Golden, "Diversity and Variability in Women's Sexual Identities," in *Same Sex: Debating the Ethics, Science and Culture of Homosexuality*, ed. John Corvino (Lanham, MD: Rowman and Littlefield, 1997).
38. Stein, *The Mismeasure of Desire*, 261.
39. *Id.*, 273.
40. *Id.*, 260. See also Charles Silverstein, "Psychological and Medical Treatments of Homosexuality," and Douglas C. Haldeman, "Sexual Orientation Conversion Therapy for Gay Men and Lesbians: A Scientific Examination," in *Homosexuality: Research Implications for Public Policy*, ed. John C. Gonsiorek and James D. Weinrich (Newbury Park, CA: Sage, 1991); Douglas C. Haldeman, "The Practice and Ethics of Sexual Orientation Conversion Therapy," *Journal of Consulting and Clinical Psychology* 62:2 (April 1994): 221-227. However,

Warren Throckmorton, "Efforts to Modify Sexual Orientation: A Review of Outcome Literature," *Journal of Mental Health Counseling* 20:4 (October 1998): 283-304 concludes, after a review of the efforts to modify sexual arousal from psychoanalytic, behavioral, cognitive, group, and religious perspectives, that some techniques to assist homosexually-oriented individuals willing to modify their patterns of sexual arousal have been effective. See also Erica Goode, "Study Says Gays Can Shift Sexual Orientation," *New York Times*, May 9, 2001, A23; Shankar Vedantam, "Studies on Gays Yield Conflicting Conclusions," *New York Times*, May 9, 2001, A13; Robert L. Spitzer, "Psychiatry and Homosexuality," *Wall Street Journal*, May 23, 2001, A26.

41. Rabbi Barry Freundel, "An Orthodox View: Homosexuality and Halakhic Judaism," *Moment* 18:3 (June 1993): 40, 43-45, at 44.

42. Roth, "Homosexuality," 126.

43. Elliot N. Dorff, *Matters of Life and Death*, 145. See also Harold Schulweis, "Morality, Legality and Homosexuals," *Jewish Spectator* 57:3 (Winter 1992-93): 23-27.

44. 15 Vermont Statutes Annotated (V.S.A.) Chapter 23 (§§1201-1207), 15 V.S.A. Chapter 25 (§§ 1301-1306), 18 V.S.A. Chapter 106 (§§ 5160-5169). Denmark enacted the Registered Partnership Act in 1989, providing same-sex couples nearly all the legal rights of heterosexual couples, followed by Norway, Sweden, and Iceland in 1993, 1994, and 1996, respectively. Leslie Goransson, "International Trends in Same-Sex Marriage," in *On the Road to Same-Sex Marriage: A Supportive Guide to Psychological, Political, and Legal Issues*, ed. Robert P. Cabaj and David W. Purcell (San Francisco: Josey-Bass, 1998). Effective in 2001 The Netherlands approved full marriage rights for same sex couples. "Dutch Legislators Approve Full Marriage Rights for Gays," New York Times, September 13, 2000, A4; "Same-Sex Dutch Couples Gain Marriage and Adoption Rights," New York Times, December 20, 2000, A6; Keith B. Richburg, "Gay Partners Make It Official," *Washington Post*, April 1, 2001, A21.

45. Bradley Shavit Artson, "Enfranchising the Monogamous Homosexual: A Legal Possibility, A Moral Imperative," *S'Vara: A Journal of Philosophy and Judaism* 3:1 (Winter 1993): 15-26, at 25.

Chapter 5

Who Is a Jew?

Before considering assisted reproduction techniques and human cloning, respectively, in Chapters 6 and 7, that raise, in part, questions of lineage, we first need to answer the question: "Who Is a Jew?" The definition of who is and who is not a Jew currently serves as a dividing line -- a boundary -- between the different Jewish movements, at least in the United States. This chapter traces how rabbinic tradition came to conclude that children of a Jewish mother are Jewish, regardless of what they do or do not believe or practice. Thus, a child's religious status follows that of his or her mother. Apart from formal conversion to Judaism, in the eyes of traditionalists, religious status follows from a biological dimension fixed at birth.

The Reform and Reconstructionist Movements have recognized the offspring of a mixed marriage, where one parent is Jewish, as Jewish, provided they are raised as Jews. Apart from conversion, the Orthodox and Conservative branches of Judaism continue to adhere to the traditional matrilineal standard, recognizing as Jews only those whose mothers are Jewish.

As background for the transformation brought about by the Reform and Reconstructionist Movements, this chapter also considers the traditional prohibition of intermarriage and the contemporary realities. Rabbinic Judaism, vigilant to combat any deterioration of the Jewish gene pool, sought to separate Jews and non-Jews, thereby avoiding contamination, reinforcing tribal loyalty, and promoting group survival and Jewish identity.

The chapter concludes with a discussion of conversion, placing it within the context of traditional norms and offering the approach of Spiritual

Judaism to both the determination of Jewish status and conversion.

Religious Status in the Jewish Tradition

Jews, among other religious, consider lineage significant. The ancients sought to maintain family blood lines. The Hebrew Bible for example, in Genesis Chapters 5, 10 and 11:10-32 and I Chronicals 1:1-9:44, records long genealogical lines. In the time of Ezra, about 450-440 B.C.E., and subsequently, an effort was made to maintain precise records and guarantee pure lines of descent.

Over the ages, the establishment of Jewish status underwent considerable transformation, moving gradually from a patrilineal to a matrilineal principle.[1] Prior to the promulgation of the Mishnah, a legal code of the third century C.E., one's father established Jewish status. Numerous genealogical tables in the Hebrew Bible attest to descent through the paternal line as the mode for determining the status of children. The legendary Jewish matriarchs, Rebekah, Leah, and Rachel, derived their suitability as wives for Isaac and Jacob through their fathers (Genesis 24:24, 47, 25:20, 29:16). Joseph and Moses married, respectively, an Egyptian and a Midianite (Genesis 41:45, Exodus 2:16, 21). King Solomon married numerous non-Jewish women (I Kings 11:1-3).

In ancient times, according to custom, women who married out of their clans went with husbands and became part of their spouse's people. Thus, in Jewish inter-tribal marriages, paternal descent was decisive (Babylonian Talmud Bava Bathra 109b).

During the Babylonian exile of the Jews in the sixth century B.C.E., intermarriage became more common. Although not instituting the matrilineal principle, Ezra, who forced the Jews to accept the authority of the Torah, cracked down on mixed marriages when the exiles returned from Babylonia in the fifth century B.C.E. He ordered Jews not to let their sons or daughters marry non-Jews (Ezra 9:12). He also urged Jewish men to get rid their non-Jewish wives and the offspring of these unions (Ezra 10:3, 11). Why? For Ezra, marriages between Jewish men and non-Jewish women had "negative" consequences. Their children were not Jewish. Also, they would not be raised as Jews. In Nehemiah 13:23-25, we read: "Also, in those days I [Nehemiah, Ezra's colleague] saw that many Jewish men had married women from Ashdod, Ammon and Moab.

Half their children spoke the language of Ashdod or of other people and could not speak the language of the Jews. I argued with them and reviled them. I beat them and pulled out their hair. I made them swear in the name of God that they or their children would not intermarry with foreigners." Ezra, however, ignored marriages between Jewish women and non-Jewish men because their offspring were regarded as Jewish.

In the centuries following the return from the Babylonian exile, Jews turned to rigid group inbreeding to safeguard Jewish identity. The authors of the Book of Deuteronomy (7:3-4) inserted the following prohibition in the Torah, providing it a Divine authority: "You must not intermarry with [the inhabitants of the land God brings you to enter and to occupy]; neither giving your daughters to their sons nor taking their daughters for your sons. If you do, they will draw your children away from God and make them worship other gods."

The third century C.E. Mishnah penalized both Jewish men and women for intermarrying. It established the rule that if a Jewish man married out of Judaism, his children would not be Jewish. However, these children could convert and become Jewish. A harsher rule applied to the children of Jewish women who married non-Jewish men. Although these children would be Jewish, they would be children of a forbidden union, similar to the offspring of an incestuous or adulterous relationship, who could never be legitimized. They would be *mamzerim* in Hebrew (Jerusalem Talmud Kiddushin 3:12). In the Jewish tradition, a *mamzer* could marry only another *mamzer*. There was no way a *mamzer* could escape this status. It was a stigma that would go on, generation after generation.

The sixth century C.E. Babylonian Talmud changed the rule that applied to Jewish women who married out of Judaism. Because a child acquired his or her mother's religious status, intermarrying Jewish women's children not only would be Jewish but also would not acquire the status of *mamzerim* or any other disabling category. However, intermarrying Jewish men's children would be non-Jewish.

In establishing the principle of matrilineal descent, a biological dimension fixed at birth, the Babylonian Talmud stated: Your son who comes from an Israelite woman is called your son [Jewish]; your son from non-Israelite woman is not called your son, but rather her son [non-Jewish] (Kiddushin 68b, Yebomot 23a).

Several reasons may account for the shift to matrilineality some fifteen hundred years ago in cases of a mixed marriages. In the first several centuries of the Common Era, Roman influence in Palestine was strong.

Roman law may have also impacted rabbinic law. Under Roman law, if a Roman male married a non-Roman woman, their children would not be Roman citizens.[2]

Matrilineal descent is often said to offer certainty in defining a child's religion because the birth mother's identity is clear. It solves the problem of children who were the product of the rape of Jewish women or where paternity could not be established. However, the possibility of uncertain paternity is no greater in mixed marriages than if both spouses are Jewish.

Despite this change in the halakhic system, even now in the rabbinic tradition, the patrilineal principle continues to have some impact. We are named after our fathers. The male line remains dominant in the descent of the priesthood. Today, among Orthodox and Conservative Jews, the offspring of a male Cohen (a priest) who marries a Levite (a priestly assistant) or Israelite (other male Jews) is considered a Cohen. Conversely, the child of a male Israelite who marries a Cohen or Levite is deemed an Israelite.

Nowadays, under the matrilineal principle of the halakhic system, the disabilities Jewish men face in a mixed marriage with non-Jewish women do not reflect shifts in family responsibilities. We cannot automatically assume that the mother, not the father, assumes primary responsibility for child rearing. In traditional two-parent, heterosexual families, mothers often exert a dominant influence during a child's formative years. But not always. Fathers are increasingly involved in child raising. According to Jewish law, a father is responsible for teaching Torah to his sons (Babylonian Talmud Kiddushin 29a). Also, with most children taking their fathers' surnames, fathers may more powerfully determine their offspring's social identities than their mothers.

Jewish Status in the Context of the Rising Tide of Intermarriage

In addition to the new roles fathers have assumed in child rearing, the rising tide of intermarriage forms the contemporary background for the attempt to resolve the age old question: "Who is a Jew?"

As a result of millennia of special efforts to survive as a small minority group in the Diaspora, the bonds of Jewish tribal loyalty remained strong. Over the centuries, social segregation and mandatory inbreeding promoted Jewish identity and group survival. As part of their strategy to shield Jews from contact with non-Jews, the rabbis promoted segregated reproduction.

Today, traditionalists, including Orthodox and Conservative rabbis, oppose intermarriage as being against the Jewish tradition and a threat to the future of the Jewish community. They want to preserve Jewish identity and lineage. Orthodox and Conservative rabbis will not officiate at intermarriage ceremonies. By officiating at a marriage of a Jew to an unconverted non-Jew, a Conservative rabbi forfeits his or her membership in The Rabbinical Assembly, the professional organization for Conservative rabbis.[3] Since the days of the Babylonian Talmud (Avodah Zarah 36b; Kiddushin 68b), the Jewish tradition has prohibited marriage between a Jew and an unconverted non-Jew based on a biblical ordinance (Deuteronomy 7:3), quoted earlier in this chapter. Although this verse refers to intermarriage with those from seven non-Jewish nations, the prohibition applies to all gentiles because after the negative prohibition, "You shall not intermarry with them," the passage continues, "For they will turn your children from Me [God] to worship other gods...." (Deuteronomy 7:4), which serves "to include all women who would turn their husbands and children away" (Babylonian Talmud Avodah Zarah 36b).

In addition to these verses continued in the Book of Deuteronomy (see also Exodus 34:16), the author of the Book of Kings warned Jewish men against taking non-Jewish wives: "You shall not enter into marriage with them, neither shall they with you, for surely they will turn away your heart to follow their gods" (I Kings 11:2). The prohibition against marrying non-Jewish men and women is also explicitly stated in the Book of Nehemiah, written in the mid-fifth century B.C.E. after the return from the Babylonian exile: "We will not give our daughters in marriage with foreigners living in our land, or take their daughters for our sons" (Nehemiah 10:31, 13:25).

Marriages of Jews with non-Jews were prohibited by Maimonides,[4] who subjected the offending parties to punishment by whipping.[5] The Hebrew Bible records that this type of sinner was also thought subject to Divine punishment (Joshua 23:13; Malachi 2:11-12).

Fast-forwarding from ancient and medieval sources to modern times, we can project that the Jewish people of twenty-first century America will be different from Jews of the past. Except for an Orthodox minority, mainly the hasidism living in isolation from the world around themselves, Jews in the United States live in an open society. One aspect of openness, namely intermarriage, will lessen Jewish identity to the consternation of those seeking to maintain the longstanding rabbinic tradition of social segregation and rejection of the non-Jewish world.

As a result of the changing social circumstances in which most contemporary American Jews live, mixed marriages engulf the Jewish community in the United States. Intermarriage is now a reality, a matter of common occurrence among Jews. A majority of American Jews who marry now find spouses who are not Jewish and only a small fraction of these non-Jewish spouses, perhaps as low as one out of twenty, convert to Judaism. Only slightly more than one-quarter of the children of these interfaith unions are raised as Jews and far fewer marry Jews.[6]

Many Jews who intermarry do not desire to abandon their Jewish heritage. Also, living in an open society we realize that non-Jews are not a forbidden species; they are not contaminants. Jewish group survival cannot be founded on the rejection of non-Jews, particularly the non-Jewish spouse of an intermarriage.

Although the Reform Movement continues to deplore mixed marriages, it remains ambivalent on this issue, looking negatively at Reform rabbis officiating at interfaith wedding ceremonies but welcoming intermarried couples into congregations. After recalling its stand, adopted in 1909, that mixed marriage is contrary to the Jewish tradition and should be discouraged,[7] the Central Conference of American Rabbis in 1973 passed a resolution declaring its opposition to participation by Reform rabbis in any ceremony solemnizing this type of marriage.[8] Likewise, the Reconstructionist Movement reserves the traditional rites of a Jewish wedding ceremony for the marriage of two Jews.[9]

To the extent that the intermarriage taboo began to collapse, liberal Judaism felt the need to respond and establish the Jewish identity of children of mixed marriages. Concern over the spiraling rate of intermarriage in the United States, the status of children of these unions, and the survival of the post-Holocaust Jewish community motivated a reinterpretation of Jewish law.

Redefining Jewish Status: The Reform and Reconstructionist Movements Break Ranks

In mixed marriage situations, the Reform and Reconstructionist Movements look to equilineal -- or nonlineal -- descent, eliminating the exclusive emphasis on the mother's Jewishness as the determining factor for a child's religious status. A father's religion thus receives a vote in determining his offspring's Jewish status. By treating men and women equally, it is fairer. Both the Reform and Reconstructionist Movements want a child with one non-Jewish parent to be raised as a Jew and to be accepted as Jewish. The equilineal or nonlineal principle makes it easier for children of a mixed marriage to be Jewish.

As early as 1947, the Central Conference of American Rabbis welcomed a child of a mixed marriage, specifically, the child of a Jewish father and a non-Jewish mother, into Reform religious schools, in which the education received was deemed "to be the full equivalent of any ritual or ceremony of conversion."[10]

In a mixed marriage context, according to both Reconstructionist and Reform authorities, having one Jewish parent does not by itself give a child Jewish identity. The Reconstructionist Movement took the lead in 1968 by adopting a resolution requiring children born of a non-Jewish mother to meet additional qualifications, including education and various religious rituals. The Federation of Reconstructionist Congregations and Fellowships (FRCF) approved the following resolution:

> We should further inform the parents [of children born of a Jewish father and a non-Jewish mother] that the Reconstructionist Movement and its affiliated institutions will consider these children Jews if the parents have committed themselves to rear their children as Jews by providing circumcision for boys, Jewish education for boys and girls and if the children fulfill the requirements of bar [or] bat mitzvah or confirmation.[11]

In 1984, the FRCF reaffirmed its 1968 resolution and required a Jewish upbringing for any child of a mixed marriage, regardless of which spouse is Jewish. The resolution identified "as Jewish a child whose father or mother is Jewish, provided that the child is raised and educated as a Jew...."[12]

The 1979 Guidelines for Conversion of the Reconstructionist Rabbinical Association (RRA), the professional organization for Reconstructionist rabbis, provide that the RRA's normal conversion procedures do not apply to infants with one Jewish parent, if both parents demonstrate their intention to raise the child as a Jew. The Guidelines state: "If one parent is Jewish, either mother or father, the offspring is to be regarded as Jewish and shall undergo the rites prescribed by our tradition (*brit milah* for boys, or a covenantal naming ceremony for girls);

but no special conversion procedure is required."[13]

Following the lead of the Reconstructionist Movement, in 1983 Reform rabbis, by a three-to-one margin, overturned centuries of traditional Jewish law. The majority supported a resolution giving a presumption of Jewish descent to any child of a mixed marriage, whether the Jewish parent was the mother or the father, provided the child is raised as a Jew, with Jewish status:

> to be established through appropriate and timely public and formal acts of identification with the Jewish faith and people. The performance of these *mitzvot* serves to commit those who participate in them, both parents and child, to Jewish life.
>
> Depending on circumstances, *mitzvot* leading toward a positive and exclusive Jewish identity including entry into the covenant, acquisition of a Hebrew name, *Torah* study, Bar/Bat Mitzvah, and *Kabbalat Torah* (Confirmation). For those beyond childhood claiming Jewish identity, other public acts or declarations may be added or substituted after consultation with their rabbi.[14]

The 1983 Resolution acknowledged a widespread phenomenon. Many children born to Jewish fathers and non-Jewish mothers were enrolled in part-time, religious programs sponsored by Reform congregations. They were being educated and reared as Jews. Recognizing these families as Jewish was consistent with the Reform Movement's concept of inclusivity and its attempt at outreach. Acceptance of the equilineal descent principle serves as a broadening factor, encouraging more families to live Jewish lives and to be involved in Reform congregations, thereby increasing temple membership and helping further the long-term survival of Jews in North America.

Two aspects of the 1983 Resolution are noteworthy. First, it establishes the Jewish status of children of mixed marriage only for Reform Jewish communities in North America, not for all Jews everywhere.

Second, a careful reading of the 1983 Resolution indicates it establishes a presumption of Jewish identity. The Resolution does not automatically confer Jewish status. It says that a child of a mixed married is Jewish only if certain conditions are met. The Resolution requires a visible and firm commitment to Judaism through meaningful acts of identification, a conscientious Jewish upbringing, including enrollment in and successful completion of a congregation's religious school curriculum and a *Bar/Bat Mitzvah*. It requires an expression of real faith and practice, not the mere accident of birth.

The 1983 Resolution raises two types of interpretative questions, some pertaining to children and others to adults. In contrast to the situation in which the non-Jewish parent agrees that the couple's home will be Jewish and all of their children will be raised Jewish, a child raised in a dual-religious environment faces confusion, in the eyes of Reform authorities, as to his or her true religious identity. For example, if the son of a Jewish father and a non-Jewish mother is being raised in a dual religion household, where their daughter is being brought up as a non-Jew, the traditional limits of patrilineality come into focus. Divided equally and intentionally between Jews and non-Jews, this type of family determined that its home is not solely Jewish but one that is both Jewish and non-Jewish with Christian (or other religious) holidays and observances prominent in the home as family celebrations. One responsum concluded, "It is far from obvious that the 'Jewish' child of such a couple, no matter what the parents' stated intent, will develop a true Jewish religious identification. We question, therefore, whether such a family is capable of transmitting an exclusive Jewish identity to one of its children."[15] Thus, the Jewish status of this child could only be established by conversion, not by an intent to raise him as a Jew or even if he undertakes "appropriate and timely public and formal acts" of Jewish identification. The Reform authorities indicated that the rabbi in the individual case must make a determination of the child's readiness to chose Judaism with the conversion occurring prior to Bar Mitzvah.

As to an adult claiming Jewish identity on the basis of his or her Jewish parent, the 1983 Reform Resolution indicates that "other public acts or declarations may be added or substituted after consultation" with a rabbi, with particular emphasis placed on "active participation in Jewish life." Reform responsa provide some guidance in this area, especially as to what indications are insufficient. One Reform responsum dealt with a man of a mixed marriage, whose father was Jewish and his mother, non-Jewish, both of whom were deceased, who wished to marry a Jewish woman. Because he could produce no evidence that he was ritually circumcised and had a Jewish education, nor been involved in any subsequent acts affirming his Jewish identity, the responsum required conversion as the affirmative means of establishing his Jewish identity.[16]

Leaving it to individual Reform rabbis to decide for themselves, on a case-by-case basis, whom they consider a Jew, the Reform authorities have eschewed establishing movement-wide guidelines or markers for what constitute the "appropriate and timely public and formal acts" marking Jewish status. A considerable degree of flexibility exists depending on individual circumstances. Some Reform rabbis may require a formal conversion before officiating a *bar/bat mitzvah* or wedding of a child of a mixed marriage, raising the specter of rifts among Reform rabbis and the emotional damage to those considered Jews by some Reform rabbis and non-Jews by others.

The Conservative Movement Upholds Tradition

Although the Conservative Movement's approach to the role of women in worship services has evolved markedly to gender equality in recent decades, for example, allowing women to be counted in a *minyan*, it has stood firm on the principle of matrilineal descent. After the shifts by the Reform and Reconstructionist Movements, The Rabbinical Assembly's Committee on Jewish Law and Standards declared in 1985 that acceptance of patrilineals as Jews by Conservative rabbis represents a violation "of a Standard of Rabbinic Practice and [is] inconsistent with membership in the Rabbinical Assembly."[17] By treating patrilineal offspring as a Jews, a Conservative rabbi forfeits his or her membership in The Rabbinical Assembly. This resolution was formally adopted in 1986 by a divided Rabbinical Assembly, marking the refusal by the Conservative Movement to overturn a rule in force for nearly fifteen hundred years. Centuries of legal precedent were viewed as compelling on this issue. Conservative authorities may have also felt that maintaining the matrilineal standard serves as a disincentive to intermarriage.

Conversion in the Jewish Tradition

The struggle over and the divisions within Judaism engendered by the question "Who is a Jew?" necessitates a brief examination of the topic of conversions.[18] Potential converts include those wanting to be married by a rabbi or those already with a Jewish spouse, who may want a clear path to the acceptance of their children as Jews. Other potential converts include adults attracted to Jewish identity. They like Jews and want to be associated with them. They may be fascinated by the intellectual power and worldly achievements of Jews.

In Biblical times, males who wished to convert to Judaism needed to be circumcised (Genesis 34:14). Females who wished to become Jews did so by marriage.

The rabbinic tradition requires a prospective male convert to submit himself to: first, circumcision or having a drop of blood drawn from his penis to symbolize circumcision if already circumcised and second, immersion in a ritual bath, a *mikvah* (Babylonian Talmud Yevamot 46a-b, Avodah Zarah 59a). Immersion in a *mikvah* is obligatory for female converts (Babylonian Talmud Yevamot 47b). Maimonides mandated adherence to these rituals.[19]

Today, even among liberal rabbis, observance of these rituals has assumed renewed importance. Circumcision, actual or symbolic, is regarded as a male convert's acceptance of the alleged covenant between God and the Jewish people. An immersion ritual, involving purification by water, is viewed as changing an individual's status. Also, by Jewish custom, not law, converts traditionally receive a Hebrew name. Male converts are generally named, the son of Abraham; females, are often named, the daughter of Ruth.

In addition to these rituals, a preparatory period of careful instruction and study as well as inner contemplation precedes the public ceremony, conducted before a court consisting of three rabbis (if this is not possible, at least one rabbi and two adult Jewish witnesses, traditionally males, but in today's egalitarian era, men or women). Both the rabbinic tradition (Babylonian Talmud Yevamot 47a-b) and Maimonides[20] require study periods. Today, a rabbi determines the requisite length of preparation, depending on the time required to obtain the necessary understanding and appreciation of Judaism so that a prospective convert may make an informed decision. Active participation in worship services and home observances reinforce the period of study.

From the point of view of various Jewish denominations, the acceptance of one's converts by other branches of Judaism remains contentious. Orthodox rabbis will not accept conversions by liberal rabbis. Conservative rabbis will, however, accept conversions by Reform rabbis, provided the conversions are performed in accordance with the Jewish tradition, at least as understood by the Conservative Movement.[21]

Spiritual Judaism's Approach to the Determination of Jewish Status and the Conversion Process

Social circumstances, notably, the high rate of intermarriage, require a new approach to answering the age old question: "Who is a Jew?"

Eschewing matrilineal, patrilineal, or equilineal principles, Spiritual Judaism, as a universalistic religion, recommends turning away from any focus on one's lineage -- maternal or paternal. We ought not strive to maintain family lineage, based on any link to or continuity of the Jewish gene pool. Jewish ancestors were not "better" or "smarter" than others.

It is who you think and feel you are, how you live your life, with whom you identify, not whose genes you are carrying, your understanding of Judaism (often requiring a period of study of up to a year or more for prospective converts), nor your undergoing a prescribed ritual or ceremony. For Spiritual Judaism, Jewish status depends on an expression of real faith in God and consistent, sincere practice, to whatever degree as individual deems important, of certain virtues, not a genetic accident, one's parents or upbringing, or undergoing a conversion ritual, symbolically changing one's identity.

Requiring an adult male convert to undergo the physical pain of an actual or a token circumcision sends the message that ritual requirements serve as a key test for Jewish status. It is also not necessary to have other, formal procedures, for instance, immersion in a *mikvah* for admission to Judaism. Although some converts may want a formal, public ceremony recognizing their acceptance of Judaism by choice, others will find a formal ceremony unnecessary.

If someone wants to have ties to the Jewish community, to have a bar/bat mitzvah as a rite of passage or be married by a rabbi, why probe his or her lineage, Jewish education and upbringing, or observance of Jewish rituals? Spiritual Judaism recommends that anyone declaring in good faith that he or she is a Jew, agreeing to worship one God, and wishing to live by certain virtues, ought to be considered a Jew, provided he or she does not accept Jesus as savior and thus is not a Messianic Jew (or for that matter, Mohammed as the last prophet).

For the twenty-first century, Spiritual Judaism looks back to and is guided by the Reform Movement's innovations in the late nineteenth century. In its 1893 Declaration, the CCAR made unnecessary the initiatory rituals of circumcision and immersion in a *mikvah* or any other ceremony or observance. The Reform Movement held it sufficient for a prospective convert to declare, orally and writing, in the presence of a rabbi and at least two witnesses, his or her acceptance of the Jewish faith, including the worship of the "One, Sole, and Eternal God." The CCAR Declaration merely required that a prospective convert: "be sufficiently acquainted with the faith, doctrine, and canon of Israel; that nothing derogatory to such person's moral and mental character is suspected; that it is his or her free will and choice to embrace the cause of Judaism...."[22]

As developed in Chapter 2, Spiritual Judaism represents a nondogmatic, nonlegalistic, path -- a liberal path, not centered on rules and rituals. It is an attempt to transcend the paradigm of rabbinic Judaism which lies at the heart of the Jewish tradition for the past 1900 years.

For each of us, including the offspring of a Jewish mother or a Jewish father (or two non-Jewish parents) or a prospective convert, Spiritual Judaism offers meaning to our existence. It provides a source of comfort enabling us to meet and surmount our daily problems and the crises we all face in living -- the hard days and the long nights. It helps us grapple with real life -- difficult people, interpersonal conflicts, failures at work, and stress that lies at the root of many illnesses -- and the immense quandaries presented by serious illness, our own or that of a loved one, as well as death and dying.

From the perspective of Spiritual Judaism, the spiritual essence of Judaism is threefold: God; interpersonal virtues; and personal virtues.[23] First, we need to realize and deep down grasp the notion of God, as the Transcendent Reality, giving order and meaning to the cosmos. In the midst of our stress-filled lives and the perplexing and seemingly overwhelming dilemmas we face, each of us can appeal to God, through prayer, meditation, and visualization, for aid and guidance in our times of need and distress for our personal and interpersonal growth and development and for our healing and inner peace of mind -- a wholeness of mind-body-spirit.

Second, our faith must influence our daily dealings with and our conduct toward others by inspiring and directing us to certain interpersonal virtues. Three are noteworthy: lovingkindness, forgiveness, and truthfulness.

Third, Judaism must help us develop our inner lives through the cultivation of certain personal virtues and character traits. Spiritual Judaism provides guideposts for daily living: helping us conquer our tendency to pride and ostentation, lessening our envy and pessimism, and curbing our anger. It points the way to a life marked by humility and self-esteem, joy and optimism, as well as equanimity and inner peace of mind.

Spiritual Judaism hopes that all Jews -- by birth or by choice -- grasp these principles and strive to apply them in their daily lives.

Let's turn and see how questions of lineage and Jewish status play out in the context of assisted reproductive technologies. In an era of mix-and-match designer children -- with one woman gestating another's egg, fertilized by the sperm of a man she never met but whom she selected for his genetic contributions to her child's potential success -- who are the mother and the father?

Notes

1. I have drawn on Shaye J.D. Cohen, "The Matrilineal Principle in Historical Perspective," *Conservative Judaism* 34:2 (Winter 1985):5-13; Phillip Hiat and Bernard M. Zlotowitz, "Biblical and Rabbinic Sources Patrilineal Descent," *Journal of Reform Judaism* 30:1 (Winter 1983): 43-48.
2. John Crook, *Law and Life of Rome* (Ithaca, N.Y.: Cornell University Press, 1967), 40-41.
3. *Proceedings of the Committee on Jewish Law and Standards of the Conservative Movement: 1986-1990* (New York: The Rabbinical Assembly, 2001), iv.
4. *The Code of Maimonides*, Book Five, The Book of Holiness, ed. Louis I. Rabinowitz and Philip Grossman (New Haven: Yale University Press, 1965), Treatise I, Laws Concerning Forbidden Intercourse, Chapter 12, 80-86.
5. *Id.*, 80.
6. Lawrence Grossman, "Jewish Communal Affairs" in *American Jewish Year Book 1993*, ed. David Singer (New York: The American Jewish Committee, 1993), 179.
7. Quoted in Central Conference of American Rabbis (CCAR) Responsum 146. "Reform Judaism and Mixed Marriage (1980)," in *American Reform Responsa: Collected Responsa of the Central Conference of American Rabbis 1889-1983*, ed. Walter Jacob (New York: CCAR, 1983), 460.
8. CCAR, *Yearbook*, Volume 83 (Philadelphia: CCAR, 1973), 97 quoted in CCAR Responsum No. 146. "Reform Judaism and Mixed Marriage (1980)" in *American Reform Responsa*, 445, 463, 464.
9. Reconstructionist Rabbinical Association (RRA), *Guidelines on Intermarriage* (Wyncote, PA: RRA, n.d.), 5-6.
10. Cited in CCAR Responsum No. 14. "Who is a Jew?" in Solomon Freehof, *Recent Reform Responsa* (Cincinnati: Hebrew Union College Press 1963), p. 76.

11. "Resolution Regarding Children of Mixed Marriage," *Reconstructionist* 34:8 (May 31, 1968): 30.
12. Resolution on Intermarriage, Passed at the Federation of Reconstructionist Congregations and Havurot 24th Annual Convention, June 16, 1984.
13. RRA, "Guidelines on Conversion," Appendix A to *Guidelines on Intermarriage*, 10-11.
14. Report of the Committee on Patrilineal Descent on the Status of Children of Mixed Marriages, Adopted by the Central Conference of American Rabbis, 94th Annual Convention, March 15, 1983, in *American Reform Responsa*, 547-550; CCAR Responsum No. 38. "Patrilineal and Matrilineal Descent (1983)," in Walter Jacob, *Contemporary American Reform Responsa* (New York: Central Conference of American Rabbis, 1987), 68. See also CCAR Responsum No. 58. "Intermarried Russian Jewish Family (1984)" in *Contemporary American Reform Responsa*, 94-95, requiring the children to perform acts of Jewish identification before the Reform Movement will consider them Jewish.
15. CCAR Responsum 5755.17. "The Dual Religion Family and Patrilineal Descent," in *Teshuvot for the Nineties: Reform Judaism's Answers to Today's Dilemmas*, ed. W. Gunther Plaut and Mark Washofsky (New York: CCAR, 1997) 251-258, at 255. See also CCAR Responsum 5754.3 "Child Raised in Two Religions," *id.*, 263-264, concluding that two sons of a Jewish father and a non-Jewish mother were not eligible to participate in a bar mitzvah because the question of religious identity had not been resolved. Although receiving informal Jewish education on a regular basis and attending synagogue services with their father at regular intervals, on Sundays they went with their mother to church school and services. CCAR Responsum No. 61. "A Child Raised in Two Religious Traditions (1983)," in *Contemporary American Reform Responsa*, 98-99, indicated that a child raised in a dual religious path could not be prepared for a bar mitzvah when at the same time he was being prepared for first communion. See also CCAR Responsum No. 131. "Assuring the Status of a Child (1989)" in Walter Jacob, *Questions and Reform Jewish Answers: New American Reform Responsa* (New York: CCAR, 1992), 205.
16. CCAR Responsum No. 39. "Patrilineal Descent and a Questionable Background (1983)" in *Contemporary American Reform Responsa*, 68-69. See also CCAR Responsum No. 59. "Three Generations of Mixed Marriage (1983)," *id.*, 95-96 requiring a conversion in the absence of Jewish education or contact. In CCAR Responsum 5753.13. "Conversion of a 'Matrilineal' Jew," in *Teshuvot for the Nineties*, 259-261, a woman, whose mother was raised as a Jew but converted to Christianity when she married, and who herself was baptized and raised as a Christian, wanted to marry a Jewish man. Requiring conversion before being accepted as a Jew, the responsum based its conclusion of the woman never previously identifying herself as a Jew and forfeiting her claim to Jewish status by way of descent.
17. Rabbis Joel Roth and Akiba Lubow, "A Standard of Rabbinic Practice

Regarding Determination of Jewish Identity," in *Proceedings of the Committee on Jewish Law and Standards of the Conservative Movement: 1980-1985* (New York: The Rabbinical Assembly, 1988), 178.

18. Shaye J.D. Cohen, "Conversion to Judaism in Historical Perspective: From Biblical Israel to Postbiblical Judaism," *Conservative Judaism* 36:4 (Summer 1983): 31-45, provides a useful historical overview.

19. *The Code of Maimonides*, Book Five, The Book of Holiness, Laws Concerning Forbidden Intercourse, Chapter 13:4, 87 and Chapter 14:5, 93.

20. *Id.*, Chapter 14:2-4, 6, 92-93.

21. Resolution adopted by the Committee on Jewish Law and Standards, February 23, 1983, in *Proceedings of the Committee on Jewish Law and Standards: 1980-1985*, 77. Rabbi David Novak, "The Status of Non-Halakhic Conversions," *id.*, 84, states:

> If a Reform rabbi performs a [halakhic conversion], I accept his converts as bona fide Jews.... These objective halakhic criteria, which alone protect the purity of Jewish identity, should not be compromised in the interests of an ultimately meaningless Jewish unity. However, rabbinical experience has taught me that a Conservative rabbi can exercise compassionate tact in urging proper [immersion in a *mikvah*] in these cases. I do not tell such converts that their conversion are invalid, but rather, that they were *incomplete*....

22. CCAR, *Yearbook*, vol. III, 94-95, quoted in CCAR Responsum No. 68. "Circumcision of Adult Proselytes (1893)," *American Reform Responsa*, 236-237. However, the Reform Movement now recognizes the social, psychological, and religious values associated with the traditional initiatory rites. It is recommended that Reform rabbis in the United States acquaint prospective converts with the halakhic background and rationale for the traditional initiatory rites and afford them the opportunity to observe these rites. CCAR, *Rabbi's Manual* (New York: CCAR, 1988), 232. CCAR, "Guidelines for Rabbis Working with Prospective *Gerim*," Adopted by the CCAR, 112th Annual Convention, June 27, 2001, 10, presume a preference for the rabbi selecting the appropriate rituals. The guidelines state that conversion should include "classroom learning, experiential learning, spiritual exploration, and rabbinic counseling." *Id.*, at 3.

23. I develop these themes in *Jewish Spirituality: Revitalizing Judaism for the Twenty-First Century* (Northvale, NJ: Jason Aronson, 2000).

Chapter 6

Assisted Reproductive Techniques: New Hope for Infertile Couples and the Facilitation of Alternative Family Arrangements

About one in seven or eight married couples in the United States, with wives aged 15 to 44, excluding those surgically sterile, suffer from infertility, a lack of pregnancy after one year of unprotected intercourse.[1] Among Jews, the percentage may be even higher because childbirth is often delayed for educational and career reasons. Although some couples adopt or opt to remain childless, many turn to modern medicine for help having children and fulfilling a fundamental emotional need.

Lack of a child (or children) leaves a painful void for most couples. A longing for a child that is unfulfilled is agonizing and anxiety producing. The anguish of barren women recurs in the Hebrew Bible. Vividly expressing her grief, the legendary matriarch Rachel cries out to Jacob, "Give me children, or I shall die" (Genesis 30:1).

Both members of an infertile couple suffer greatly. They experience frustration and degradation in their inability to have children. They feel inadequate and incomplete as men or women. Guilt, shame, anger, and humiliation mark their emotional landscape. Infertile couples are often very sad and depressed. They experience overwhelming feelings of powerlessness and failure. They are embarrassed and defensive in explaining why they do not have children. Some marriages disintegrate as a result of the tension and stress engendered by unsuccessful attempts to conceive.

Infertile couples turn over their days and their lives to gynecologists, urologists, reproductive endocrinologists, and laboratory technicians.

They embark on a long and often expensive medical roller coaster of rising hopes and crushing failure that all-too-often characterizes infertility treatments, marked by years of heartbreaking disappointment. Women spend months with thermometers by their beds, dreaming of the baby they want, but may not be able to have. Monitoring a woman's body can become almost a full-time job.

For couples seeking treatments for infertility, about 85-90% can be treated using conventional medical procedures, such as surgery, hormone treatments, and fertility drugs.[2] Many of the remainder, encountering still unexplained infertility, opt for assisted reproduction technologies after spending considerable time and money on standard therapies.

Assisted reproductive technologies -- artificial insemination, egg donations, in vitro fertilization, surrogate mothers -- represent the last hope for biological offspring. These stressful, expensive, and time-consuming techniques seek to relieve the anguish of men and women who previously would have abandoned any hope of having children. Women arrange their work schedules around clinic visits, undergo numerous invasive, often painful, emotionally-charged procedures. About 50% of the infertile couples using modern techniques are successful if they persist long enough; the other half remain unsuccessful, despite various treatment modes.[3] However, there is always another procedure, another specialist.

Commercial sperm banks thrive, with some of them specializing in genetic choice, such as sperm from geniuses. The American surrogacy industry flourishes in states allowing surrogacy.

Egg donors offer a shining hope for aging baby boomers, many of whom try to conceive in their late thirties and early forties. In particular, egg donation has revolutionized the field of reproductive technology, increasing the success rates and creating a huge demand for human eggs. Ads appear in prestigious college newspapers offering payments of up to $50,000 for donors with stratospheric SAT scores and other desired attributes.[4] Infertility organizations typically pay egg donors far less than the payments offered in advertisements to compensate for a donor's time and discomfort, not as consideration for the eggs.

After considering the biblical admonition to be fruitful and multiply and the variety of assisted reproductive techniques, this chapter examines traditional and liberal approaches to each of these technologies. As we will see, significant differences exist between traditionalists and more liberal authorities, especially with respect to the use of donated sperm or eggs as well as surrogate mothers. Among traditionalists, one of the most heated halakhic debates of the twentieth century has focused on the permissibility of artificial insemination by donated sperm. The perspective of Spiritual Judaism is also considered.

"Be Fruitful and Multiply"

Judaism values children as a Divine blessing for their parents (Deuteronomy 7:12-14) and the Jewish community.[5] For traditionalists and Conservative rabbis, having children represents the fulfillment of a religious obligation, an alleged Divine commandment. According to one account of creation in the Book of Genesis (1:28), God blessed and charged humans to "be fruitful and multiply and fill up the earth." God, we are told, made this statement in the form of a commandment to Noah after the flood (Genesis 9:1, 7). Abundance of offspring is a recurring promise to the legendary patriarchs whose descendants, we are told, shall be "like the stars in the heavens and the sands of the sea" (Genesis 15:5, 17:16, 22:17, 28:14, 32:13, 35:11; Exodus 32:13; Deuteronomy 1:10, 10:22). Throughout the centuries, Jewish sages have declared, "One without children is considered as though dead" (Babylonian Talmud Nedarim 64b).

The birth of a child validates a couple's sexual partnership as a means of providing Jewish continuity. For the rabbis (Babylonian Talmud Yevamot 64a), loving relationships, marked by the birth of children, serve as a dwelling place for the Shekhinah, the female aspect of Divine Presence, on earth; conversely, a childless marriage brings about her departure. As we read, "Wherever your seed is after you, the Shekhinah dwells; wherever there is no seed after you, upon whom will the Shekhinah dwell, on wood or stones?"

According to the rabbinic tradition, failure to perpetuate the species is equivalent to the diminution of human life and is morally similar to murder. As Rabbi Eliezer stated, "He who brings no children into the world is like a murderer" (Babylonian Talmud Yevamot 63b).

Because humanity is God's image on earth, failing to reproduce deprives the world of the Divine quality, reflected in each new life. As Jewish sages put it, "[H]e who does not engage in procreation is as if he diminished the Divine image" (Babylonian Talmud Yevamot 63b). Without human descendants, no one embodies God's image. With fewer worshipers, the pool of souls would decrease thereby frustrating the fullness of creation. If humans are not fruitful, they also deny the essential purpose of creation, namely, to serve as a witness for the Holy One. The diminution of humans deprives God of human witnesses to the Divine glory.

For the Jewish sages, procreation enables homes to be prepared for souls on earth thereby accelerating the coming of the Messianic Era. According to Rav Assi, the Son of David will not come before all the souls in the afterlife Divine Storehouse, the place of each soul's origin and termination, have come into this world (Babylonian Talmud Niddah 13b).

In the mystical Jewish literature, propagation occupies a significant place. We are told in the *Zohar* (I, 12b-13a), the main book of the Jewish mystical tradition:

> The *sixth precept* is to be fruitful and multiply. For he who performs this precept causes the stream [of Divine influence] to be perennially flowing so that its waters never fail, and the sea is full on every side, and new souls are created and emerge from that tree [the symbol of the realm of the tree of life, the Divine Sefirot] and the celestial hosts are increased in company with these souls.... [H]e who refrains from propagating his kind derogates, if one might say so, from the general form in which all individual forms are comprehended, and causes that river to cease its flow and impairs the holy covenant on all sides.... [A]s for his soul, she will not enter at all "within the curtain," and will be banished from the next world.

The universe is enriched by the beneficent acts of newly created souls who stimulate the flow of Divine Energy and heavenly abundance. Conversely, when humans neglect to be fruitful, the Divine Presence departs with the Divine Energy, otherwise ready to create new life, deflected from its positive course into unproductive, even destructive channels and evil forces.

Although having children in today's egalitarian era represents a couple's *mitzvah*, Jewish law places the obligation to procreate on males, not females (Babylonian Talmud Yevamot 65b). This may reflect the context of the ancient world in which men had greater control than women over whether they would marry and thus procreate. It was a sin for men to remain unmarried and childless. However, the obligation ceased to apply to infertile males because God created some bodies this way.

Because pregnancy and childbirth involve pain and risks, women were encouraged, but not obligated, to accept these risks and jeopardize their lives.[6] It may also reflect an understanding of the emotional and social dimensions of pregnancy; women's lives change enormously as a result of childbearing and rearing.

The uncertainties of ancient life, resulting from wars, famine, and disease, required a large number of births to permit the survival of at least some offspring. A minimum of two children for each family was required. The Talmud (Babylonian Yevamot 61b) states, "One must not abstain from 'fruitfulness and increase' unless one has children. The School of Shammai says: two males, derived from Moses who had two sons. The School of Hillel says: a male and a female, as it is written, 'God created them male and a female'" (Genesis 5:2). (See also Jerusalem Talmud Yevamot 6:6). Jewish law follows Hillel's view based on God's creation of Adam and Eve,[7] but encourages continually engaging in procreation even if a couple has a son and a daughter.[8] The rabbis (Babylonian Talmud Yevamot 62b) urge couples to have as many children as possible based on Isaiah 45:18, where God, the prophet tells us, created earth as a place for people to live. Furthermore, in the Book of Ecclesiastes (11:6) we read: "Do your planting in the morning and do not hold back your hand in the evening, because you do not know which is going to succeed, the one or the other, or if both are equally good."

Assisted Reproductive Techniques: An Overview

Assisted reproductive techniques offer bold approaches to the problems of infertility or the desire for a child by single persons or homosexual couples. Three modern techniques exist: artificial sperm insemination and egg donation; in vitro fertilization; or surrogacy.

Artificial Insemination and Egg Donation

Artificial insemination involves introducing a sperm into a female's genital tract, without sexual intercourse, for the purpose of fertilization. This is the first technique used if a couple cannot conceive through intercourse because of sexual dysfunction, namely, the husband is sterile or impotent, insufficient sperm exist in a husband's semen, or incorrectly shaped or energized sperm are unable to swim up the vagina. It is also used by single women or lesbian couples.

Artificial insemination enables a woman who wants to, and is able to carry a fetus, to give birth to a child of her own. Likewise, for a woman, single, married, or part of a lesbian couple, who cannot produce ova, egg donation permits in vitro fertilization by her husband (or another male). The use of donor eggs is soaring because it is the most successful assisted reproductive technology available, especially for the in vitro fertilization of older women. Using young women's eggs, the success rate does not drop as a recipient's age increases.

Two types of artificial insemination exist. The first uses a husband's sperm and is known as artificial insemination from husband, or AIH. No third party is involved. Through AIH the offspring is genetically linked to both parents. The child may physically resemble the husband.

To overcome a husband's inability to produce healthy sperm (or for single women or lesbian couples), the second method uses a donor's sperm, artificial insemination from donor, or DI. With DI, the child is genetically linked to one parent -- his or her mother; the child inherits his or her genetic father's genes. The sperm donor influences a child's medical history, intelligence, height, physical appearance, and susceptibility to diseases.

Similarly, with egg donation, the child is genetically linked to his or her mother; the child inherits the egg donor's genes.

In Vitro Fertilization

In vitro fertilization (IVF) assists women with damaged or absent fallopian tubes, the place of fertilization and the path by which a fertilized ovum reaches the uterus. With this type of assistance, a woman is capable of carrying a pregnancy to term. IVF is also used to deal with other female infertility problems, such as ovulatory difficulties, with various male difficulties, and with unexplained infertility.

Through in vitro fertilization, a woman's egg, whether her own or donated by another female, is fertilized by sperm outside her body. The egg and the sperm are combined in laboratory equipment, usually in a petri dish containing an appropriate culture medium. Today, if a husband has a low sperm count, so that there are never enough sperm, sperm can be directly injected into an egg taken from his wife, through a procedure called intracytoplasmic sperm injection. With IVF, the resulting embryo is then transferred back into his wife's uterus (or is implanted in another woman's, a surrogate's, uterus if his wife is incapable of carrying a pregnancy) who gives birth.

This technique begins by giving woman hormones to stimulate the production of several or even dozens of eggs a month, rather than the typical one, so that more can be fertilized and transferred back giving rise to the chance of multiple fetuses, a topic considered in Chapter 9. The use of hormones to cause a woman to hyperovulate increases some risks, including ovarian cancer, stroke, and heart attack, but these risks are not too high.

Eggs are collected shortly before ovulation occurs, when a physician inserts a hollow needle into the abdomen, and aided by either an optical instrument or an ultrasound monitor, guides the long needle to gather the developed ova. After the ova have been inspected to ascertain that they are not defective, they are then combined with the prepared sperm. One or more fertilized eggs are permitted to develop in culture for up to a few days, until they reach the stage of between two and eight cells. At this point, each developing fertilized egg is called a preembryo, with the fusion of male and female genetic material achieving completion when a preembryo reaches the eight cell stage. On the third or fourth day, the preembryos are then transferred to the woman's uterus, using a small vaginal catheter inserted through her cervix. When one or more of the preembryos have attached to the uterine wall, pregnancy goes forward in the usual way to fetal development.

Two alternative IVF procedures also exist. With a gamete intrafallopian transfer (GIFT), the egg and sperm, still separate, are brought near one another and then placed by a catheter directly into the fallopian tube, where the egg and sperm normally meet, thereby replicating more closely the natural process of fertilization. Using a zygote intrafallopian transfer (ZIFT), the egg and sperm have already united outside the body and the preembryo produced in vitro is transferred to the fallopian tube, again attempting to imitate natural fertilization more closely. ZIFT has the advantage over GIFT because the early preembryonic development can be evaluated by observing the cellular division. However, because physicians generally use IVF only to circumvent absent or dysfunctional fallopian tubes, these alternative procedures are infrequently utilized.

In the future, we will likely see the in vitro maturation of eggs. Physicians will remove a small slice of a woman's ovary containing hundreds of eggs, to be frozen and matured at will. A woman will undergo this minor operation in her late teens with the eggs stored until she wants to have children.

Surrogacy

If a couple is unable to conceive or a wife (or a single woman or a lesbian couple) cannot retain a pregnancy, the couple may turn to another woman -- a surrogate -- to bear the child. A surrogate mother is either a gestational surrogate or an ovum surrogate.

The opportunity to benefit from gestational surrogacy arises when a woman, suffering from one of a number of fertility problems, including an absent or malformed uterus, a medical condition, such as diabetes, making pregnancy dangerous, or a condition endangering a fetus, may be capable of producing eggs, but cannot carry a pregnancy. The surrogate gestates or incubates the fetus. A zygote is formed by the in vitro fertilization of the genetic mother's egg, and surrogate -- the birth mother -- is impregnated with this fertilized egg. Through this type of gestational surrogacy, the offspring is genetically linked to both parents (or to a single woman, one partner of a lesbian couple and the sperm donor, or to both partners of a lesbian couple whose eggs have fused); the birth mother has no genetic link to the newborn. Thus, the mother allows her womb to serve as the fetus's incubator during gestation without contributing any genetic materials.

For married couples, there are three other possible combinations apart from the husband's sperm and the wife's egg. These are: first, the husband's sperm and a donated egg; second, the wife's egg and donated sperm; third, donated sperm and donated egg (from a woman other than the wife or surrogate mother). In option one, by using the husband's sperm, the husband is genetically related to the offspring. With options one and three, the woman, who carries the child, even if the egg came from another woman, has the satisfaction of being the gestational mother. Because surrogacy enables a couple to have a child bearing the genes of at least one of them, the last option is rarely used; adoption would serve as well or better.

If the surrogate is an ovum surrogate, both her egg and her womb are used. A wife, a single female, or one member of a lesbian couple, who lacks healthy ovaries (or the ability to produce eggs) and is unable to carry a fetus to term will opt for ovum surrogacy. The surrogate's egg is fertilized in vitro by artificial insemination with the intended father's (or a donor's) sperm. Through this type of ovum surrogacy arrangement, the offspring is genetically linked only to the husband-father. It is also possible to inseminate the surrogate mother's egg with donated sperm. Married, heterosexual couples rarely use this alternative because the child has the genes of neither intended parent. In any event, the ovum surrogate is bound to the child, genetically and gestationally.

Although the surrogate mother gives birth to the child, she delivers the newborn to the intended parents (or parent), pursuant to a surrogacy agreement signed prior to conception. In this agreement, the surrogate mother waives all parental rights to the child.

The Jewish Tradition: Artificial Insemination and Egg Donation

Donated sperm and/or eggs may be used by a married couple, a single woman or a lesbian couple. Traditionalists distinguish between AIH (artificial insemination by husband) and DI (donor insemination) with the later raising a number of complicated issues and intense halakhic debate. Donated eggs pose similar difficulties for traditionalists.

Artificial Insemination by Husband

Traditionalists generally approve of AIH where pregnancy can be achieved in no other way after a long waiting period ranging from two to eight years, provided only the husband's sperm are used.[9] As one Reform halakhic expert has noted, "...[T]he vast majority of [traditional authorities] not only permit, but encourage AIH in fulfillment of the *mitzvah* be 'fruitful and multiply.'"[10]

Many traditionalists worry about the means used to obtain the husband's sperm. These authorities voice concern about destruction of seed in vain (Babylonian Talmud Niddah 13b), based on the rabbinic interpretation of the sin of Onan, the improper emission of semen for naught (Genesis 38:8-10), discussed in Chapter 4.

In the mystical tradition, as we have seen in Chapter 4, sages relying on the *Zohar*, declare that the wasteful emission of sperm is the gravest of all sins. For the mystics, a man who masturbates, preventing the use of that semen for conceiving a child, is guilty of murdering a potential child. Even the involuntary emission of semen was said to create demons endangering not only him who masturbates but also the entire community.[11]

Despite these fears, a majority of traditionalists have concluded that AIH is not an instance of the wasting of seed.[12] Capturing semen for the purpose of procreation is not wasting it. A husband intends to procreate. His semen is put into the uterus to impregnate his wife. It is used to fulfill his responsibility for procreation. Any objections to masturbation are generally set aside because the procedure overcomes a husband's infertility difficulties.

Permissive traditionalists have devised a hierarchy of methods for collecting semen from a husband for AIH procedures.[13] Gathering sperm from the vagina following intercourse is the preferred method; however, this generally is unrealistic. Then, semen may be collected from a condom used during intercourse or a receptacle placed in the wife's vagina. If none of these methods may be used, sperm to treat infertility may be obtained by masturbation, preferably performed by a physician using a mechanical stimulator, although a few authorities allow self-stimulation.

Although traditionalists retain their abhorrence of masturbation, liberal authorities do not, ignoring the prohibition and supporting AIH. As one Reform responsum put it, "Masturbation should be discouraged, but we would not consider it harmful or sinful."[14] Through AIH a husband participates in procreation in the only way he can; otherwise, a husband faces humiliation and even depression if he cannot impregnate his wife through sexual intercourse. Conservative Rabbi Elliot N. Dorff concludes, "[W]e should do all we can to make him feel good about the process and the child that results."[15] Furthermore, by using the husband's sperm, the child is his.

Artificial Donor Insemination and Egg Donation

DI and egg donation present difficult issues for traditionalists: adultery, incest, lineage, and moral values, in addition to the problem of collecting donor sperm because of the prohibition against wasting seed just discussed.[16]

Possibility of Adultery

A minority of traditionalists view DI as adulterous because it represents the deliberate introduction of a third party's, the donor's, sperm into the vagina of a married woman even if it is done without sexual intercourse, based the mandate contained in Leviticus 18:20: "Do not have intercourse with another man's wife."[17] Likewise, they view the use of a donated egg as violating the exclusivity of a marital relationship.

In Jewish law, a married woman is sexually forbidden to all other men until her husband dies or they are divorced. Reflecting a patriarchal bias, Jewish law is more permissive for husbands. A married man does not commit adultery by having sex with an unmarried woman. Also, a single woman does not commit adultery if she has sex with a man, married or single.

Adultery committed by a wife, even with her husband's consent, represented a serious offense, namely, the destruction of the family unit. According to the Hebrew Bible (Leviticus 20:10; Deuteronomy 22:22), adultery was a capital crime, punishable by death for the woman and her lover.

Over the ages, for the rabbis, the stigma of adultery remained. After a wife's adultery, the spouses were obligated to divorce even if they wish to continue their marital relationship. Adultery was viewed not only as an offense against the husband but also as a trespass against God in violation of biblical requirements of the sanctity and purity of a marriage and family life. Furthermore, the wife could not marry her lover.

Any child born of an adulterous relationship, one between a married woman and a man not her husband, resulted in the defilement of the offspring, rendering him or her illegitimate, a *mamzer*. In Jewish law, although the stigma could not be removed, an illegitimate child was not deprived of any financial rights, including support and inheritance; however, this child was not accepted into the Jewish community. The Hebrew Bible (Deuteronomy 23:3) prohibits marriage between a legitimate and an illegitimate person, even in the tenth generation, but not between two illegitimates. The concept of a *mamzer* likely sent shudders through anyone contemplating an adulterous relationship.

We now want marital fidelity on the part of both spouses. Infidelity by either is equally unacceptable. Thus, the prohibition against adultery (Exodus 20:13; Deuteronomy 5:17) strives to protect the integrity of a marital relationship. Men and women are prohibited from having sexual relations with someone other than their spouses.

Most traditionalists, conclude that adultery requires sexual intercourse, specifically, that a man's penis enter a woman's vagina. With DI, no contact of genital organs occurs. Thus, under this majority view, DI is not adultery and the child is legitimate.[18] Also, with egg donation there is no contact of genital organs; in the case of DI or egg donation, furthermore, any adulterous intent is lacking.

Beyond the physical aspects of an extra-marital relationship, the implications flowing from the marriage bond may underpin the arguments among traditionalists over whether or not DI is adulterous.[19] For those who see DI as adulterous, marriage goes beyond a husband's exclusive rights to his wife's sexual intimacy. The marital bond also encompasses his exclusive rights to control the activation of his wife's reproductive capacity. Thus, DI severs procreation from the marital union thereby violating, it is asserted, the very essence of a marriage. DI is seen as destroying the purity and holiness of Jewish family life. As a source of division and mental suffering, not happiness, DI is seen by a minority of traditionalists DI is seen as bringing about the destruction of the family unit. A majority of traditionalists, however, view the marital bond as concerned only with sexual fidelity.

Liberal commentators reject any question of adultery. For example, Conservative Rabbi Dorff, in a responsum adopted by The Committee on Jewish Law and Standards, does not consider DI or egg donation as adulterous.[20] Adultery, for Dorff, destroys the trust and moral commitment between husband and wife serving as the foundation of their marital relationship. Jewish sages were concerned about the chaos that might result from unbridled sexual license. Possessive men and women might seek revenge on others who had slept with their marital partners; conflict and even murder might result from adulterous liaisons. Neither DI nor egg donation raises any of these risks.

With DI or egg donation, one's spouse knows about the donation. A husband and wife want to have children in the context of mutual openness and trust. There is no intent to form an illicit relationship. The couple affirm their love for each other in having and raising a child. Marital fidelity remains intact where a husband consents to DI or a wife to an egg donation.

Possibility of Incest

Traditionalists also oppose DI or egg donation because of concerns about possible, unintentional incest.[21] With an anonymous sperm donor, a child resulting from DI may inadvertently be involved in an incestuous relationship with his or her biological father's family. A child produced by DI might marry one of the children of the sperm donor and his wife, resulting in half-sibling incest between a half-brother and a half-sister. A son and daughter born of the same sperm donor and raised in separate homes may meet as strangers and marry in violation of the prohibitions contained in Leviticus 18:9 and 18:11.

According to the Book of Leviticus, the incest taboo results in a series of individuals being sexually forbidden to each other. On marriage, numerous members of one's spouse's family also become off limits. These incest prohibitions encompass sexual intercourse with one's parent (Leviticus 18:7), stepparent (18:8), one's sibling, step-or-half sibling (18:9, 11), child or grandchild (18:10), aunt or uncle (18:12, 13), aunt or uncle's spouse (18:14), child's spouse (18:15), or sibling's spouse (18:16). Over the years, the rabbis expanded the incest boundaries. Maimonides provided a good summary of the rabbinic expansion of the rules against incest.[22]

The rationale for the incest prohibition is quite simple. Incest destroys the trust between siblings or other close relatives. If incest involves two people in the same household, it may be exploitative, if one is dependent on the other. Thus, a man might take advantage of women in his household who were easily accessible to him and dependent on him for financial support. Because it involves an abusive relationship between family members of unequal status and power, incest destroys the weaker member's healthy personality, inflicting deep emotional damage that may never be healed. The incest taboo exists because of the fear of genetic abnormalities in the offspring of blood relatives.

Prohibiting incest was also viewed as making Jews holy and pure. The author of the Book of Leviticus sought to separate the Jews from other ancient peoples. God stated, "You shall not copy the practices of the land of Egypt where you dwelt, or of the land of Canaan to which I am taking you; nor shall you follow their laws. Do not have sexual intercourse with any of your relatives" (Leviticus 18:3, 6, also Leviticus 20:23, 24, 26).

There are several answers to the incest possibility with DI or egg donation. First, traditionalists do not recognize family lineage among non-Jews through the father's line. If the anonymous sperm donor is non-Jewish, sexual intercourse between a child born of his sperm donation and those born through his marriage, even if his wife is Jewish and his children are Jewish, would not violate the Jewish law against incest. The child is only related to his or her Jewish mother and thus is Jewish; he or she would not commit incest by marrying a half-sibling from a genetic, but non-Jewish father. Why? A Jewish child in this situation would have no halakhic relationship to its non-Jewish biological father. Thus, permissive traditionalists would allow only a non-Jewish man, but not a Jewish man, to donate sperm for insemination in a woman who is not his wife.

Second, even though a child resulting from sexual intercourse between an offspring produced by the donation of sperm (or egg) and the donor's other children is genetically incestuous, legally and morally it is not incestuous. The requisite intent is lacking. The couple lacks any knowledge that they are half-siblings.

Third, and perhaps most fundamentally, we should not focus on theoretical scenarios, often of the most extreme kind. According to one Reform responsum, "In our very large, widely dispersed American Jewish community," the likelihood that the child may marry and unintentionally violate the prohibition of incest, "is minimal...."[23] Thus, for liberal authorities, "...[T]he possibility of the child marrying one of his own close kin is far-fetched, but that since, according to Jewish law, the wife has committed no sin and the child is [legitimate] [because there is no forbidden sexual intercourse], then the process of [DI] should be permitted."[24]

To avoid any possibility of incest under American or Jewish law among either the genetic offspring arising from donated sperm or egg or the genetic offspring born by donation and those within a donor's own family, ideally a sperm or egg donor's identity should be known to the social parents (and the child on attaining age 18). The hard reality conflicts with the ideal. Sperm donations typically are anonymous, unlike egg donations where the donors are known, often a wife's sister or cousin. Even if a donor's identity remains confidential, the social parents need to try to learn as much as possible about the donor's medical history and personal identity and character. This information enables a child to know how to deal with genetically inherited diseases or susceptibility to such a disease. It also helps a child achieve a greater sense of identity. Regardless of state law requirements (or those of sperm banks), donors ought to supply information about their medical history and their physical and personal characteristics. If married, a donor needs to inform his or her spouse and gain his or her consent.

Lineage Concerns

Apart from the possibility of incest, the Jewish tradition is concerned about lineage on both a child's father and mother's side. Attention focuses on avoiding any pollution of the "purity" of the Jewish genetic line. Knowing the identity of one's ancestors is also said to give them immortality by advancing their aspirations.

With procreation within a family unit, the genealogy of offspring is known thereby, avoiding any ambiguity with respect to paternal identity. DI, using semen of an anonymous donor, renders identity of the DI father virtually impossible to detect. DI brings a child of unknown paternal origin into the world, blurring the child's genealogy. A fear exists that family lineage would be confused, harming a child's sense of identity and undercutting the traditional understanding of the family unit. Focusing on the severance of the link between marriage and the procreation of children, Lord Jakobovits concluded:

> Artificial insemination utilising an outside donor... is, however, considered to pose grave moral problems. Such operations, even if they may not technically constitute adultery, would completely disrupt the family relationship. Moreover, a child so conceived would be denied its birth-right to have a father and relations who can be identified. Altogether, to reduce human generation to "stud-farming" methods would be a debasement of human life, utterly repugnant to Jewish ideals and traditions....[25]

For the mystics, the idea of introducing non-Jewish sperm with its allegedly defective character possibilities, into a "pure" Jewish woman represents an abomination.[26] According to Rabbi Shlomo Halberstam, the twentieth century Hasidic Bobover Rebbe, only a child who has a ritually fit Jewish father can be considered holy and of good lineage; only this type of child inherits all the holiness inherent in the Jewish people. The use of a non-Jewish donor may surreptitiously introduce "suspect" children into the Jewish people. Through DI, impurity, he asserted, would infiltrate Jewish bedrooms. Rabbi Mayer Amsel, another twentieth century commentator, stressed the "polluted" nature of non-Jewish sperm and all the offspring generated from it. In arriving at his opposition to DI, he relied on the *Zohar's* discussion of an altercation between an Israelite man and the son, of an Israelite woman and an Egyptian man, in which the latter, the son cursed God and was executed (Leviticus 24:10-14). For Rabbi Amsel, everyone who comes from impure sperm will eventually reveal this publicly because its pollution resides eternally.

In contrast to Jews who are "pure" and "holy," some traditionalists maintain that a child produced by means of DI is "impure" and "polluted." The use of non-Jewish sperm is said to blur the distinction between the forces of "purity and good" and those of "impurity and evil."

In response to those wanting to preserve the "purity and goodness" of the Jewish gene pool, Spiritual Judaism reminds them that we are all created in God's image. We are all equal before the Eternal. For the twenty-first century, Spiritual Judaism offers an ethic of universalism, not ethnic purity.

In the United States, if a child is born through the use of DI, the mother's husband, not the sperm donor, is the child's father. American law views the social father, man who raises the child, as his or her father for all legal purposes. All statutes in the United States regulating DI (there are more than thirty state statutes, with some eighteen modeled after Section 5 of the Uniform Parentage Act[27]) provide that the offspring of the donor sperm are to be treated as the consenting husband's -- the rearing father's -- offspring for a variety of legal purposes, including legitimacy, inheritance, and support. To obtain this result, certain statutory requirements must be met: typically, the mother is married; her husband consents in writing to DI; and a licensed physician performs the procedure.

Under a contractual (non-statutory) approach to offspring status, a similar result applies provided the donor, the recipient, and the recipient's husband agree that the husband will assume all rearing rights and duties with respect to the child and that the donor will have no rearing rights and duties. In other words, by contract, a sperm donor relinquishes all parental rights in advance of conception in return for receiving consideration for his role in providing the sperm needed for procreation. With the establishment of a father-child relationship between the child and the wife's husband, by statute or contract, sperm donors are immune to financial claims for support of progeny born through the use of DI.

Only a few states, in contrast, have enacted legislation regarding egg donations.[28] These state statutes clarify that the donors have no rights or duties with respect to raising the child.

From a traditional Jewish perspective, the child born through the use of DI can use his or her social father's name for Hebrew name identification purposes, such as when called to the Torah. For Orthodox and Conservative Jews, priestly status is, however, determined by one's biological father. If a sperm donor, the genetic father, is a Cohen, Levite, or Israelite, the child has the same status. Otherwise, by default the child is treated as an Israelite.

A word of practical advice is in order with respect to sperm or egg donations as well as IVF or surrogacy situations. Parents should tell a child when he or she is old enough to understand that he or she was conceived with the help of a donor, whether, sperm, egg, and/or womb. They should not keep it a secret, especially within the family unit. Openness on the part of parents and child helps maintain family harmony as well as the emotional health and well being of all. This is easier said than done[29] and Judaism needs to develop counseling services to facilitate it.

Moral Values

Traditionalists raise concerns about a general decline in moral values pertaining to marriage and family. Approval of DI or egg donation, some maintain, may encourage sexual licentiousness and promiscuity by creating an environment where the insemination of a wife by another man's sperm (or the use of another woman's egg) is considered morally acceptable. Traditionalists are really concerned about any type of procreation outside the boundaries of sexual intercourse between a man and a woman who are married to each other. Pointing to the decoupling of the tie between sexual intercourse and parenting, Lord Jabokovits feared:

> ...the abuses to which its [DI's] legalisation would lead, however great the benefits may be in individual cases. By reducing human generation to stud-farming methods, [DI] severs the link between the procreation of children and marriage, indispensable to the maintenance of the family as the most basic and sacred unit of human society. It would enable women to satisfy their craving for children without the necessity to have homes or husbands.... Altogether, the generation of children would become arbitrary and mechanical, robbed of those mystic and intimately human qualities which

make man a partner with God in the creative propagation of the race.[30]

In the twenty-first century, we are witnessing the development of new family configurations. Chapter 4 considered the need for Judaism to look beyond the traditional, heterosexual family unit. DI makes it possible for a single, heterosexual woman or a lesbian couple who want a child to be artificially inseminated (or inseminated and implanted with another woman's egg). Egg donation coupled with use of a surrogate also enables an unmarried man or a gay couple to have a child.

Stepping back from these possibilities, let's focus on an infertile, heterosexual, married couple. Spiritual Judaism applauds the use of DI or egg donation. The Jewish tradition has always valued children. Having and raising Jewish children is a current imperative for the Jewish community which suffers from a demographic crisis, smaller families, intermarriage, and losses from the Holocaust.[31]

Compassion also enters into the equation. A childless, infertile couple is distressed. They suffer greatly. Spiritual Judaism advises let them go ahead and use DI (or donated eggs) to achieve their goal of having a child.

Human Factors

Liberal authorities raise a variety of personal and psychological queries. One Reform responsum dealing with an egg donation from the wife's first cousin with the embryo implanted in the wife's womb, raised a number of issues: will the source of the egg cause psychological difficulties; will it aggravate normal family strife; are any pressures for donation being applied; have the parties considered the possibility of a disabled child; and what will be the impact of egg donation on the couple's marriage? However, the responsum concluded: "We would give reluctant permission to use *in vitro* fertilization in the manner you have described [with the wife's first cousin donating eggs for IVF with the husband's sperm and embryo being implanted in the wife's womb]. The potential problems are numerous and should lead to great caution."[32]

Despite these misgivings, DI or egg donations generally will help heal, not harm, a family. Fears centered on personal and psychological difficulties for all involved and the potential to add strain to a marriage are overblown. These procedures typically will not lead to divisiveness, particularly envy on the part of spouse unable to provide sperm or eggs.

A husband usually will not resent the growth of a child in his wife's body.

These procedures, furthermore, will not complicate the relationship of the child to his or her rearing parents. The use of donated sperm or eggs reflects marital love. Children born of these procedures flourish in homes of warmth and love. Empirical studies indicate that the quality of parenting in families with a child conceived by assisted conception is superior to that shown by families with a naturally conceived child.[33]

A Donor's Perspective

Despite the fact that donors help bring a child into the world, many traditionalists would prohibit the donations of sperm or eggs by Jews, reflecting the concern that sperm or egg of a Jewish donor may be used by a non-Jewish recipient.[34] These authorities view the procurement of semen from a Jewish man for insemination in a non-Jewish woman as impermissible because non-coital emission of semen can be approved only for purposes fulfilling one's obligation regarding procreation, such as by AIH, but not DI.

Furthermore, traditionalists will not sanction egg donation by a Jewish woman if it results in a situation in which a half Jew-half gentile child is to be reared as a non-Jew. If the child's status is that of a non-Jew, the egg donation is viewed as contrary to the norms of Judaism. Permission would be granted for such assistance only when withholding it would result in enmity towards Jews.[35]

Liberal commentators support the donation of sperm or ova. This act helps bring children into the world. It reflects compassion toward an infertile couple who could otherwise not have a child.

Because an egg donor faces some additional risks and physical dangers in providing ova, however, she should seek medical approval that she can donate eggs without significant danger to her health. In light of the medical risks to an egg donor, the Conservative Movement concluded that this technique "should only be used when the couple has seriously considered all other options for having children, including adoption."[36]

The Jewish Tradition: In Vitro Fertilization

Most traditionalists approve of the use of IVF provided a husband's sperm and his wife's eggs are used, thus obviating any question of adultery, and lengthy waiting periods are observed, ranging upwards from two years after marriage. Despite fears about risks for the fetus and an increased likelihood of seriously impaired children, neither of which has occurred, Modern Orthodox Rabbi J. David Bleich, a bioethics expert, concluded, "Yet, if properly controlled and not permitted to become a substitute for normal human procreation...[IVF] can be a welcome means of bestowing happiness and fulfillment of parenthood upon otherwise childless couples."[37] IVF is accepted by the Conservative[38] and Reform[39] Movements.

A minority of traditionalists, however, condemn IVF on two grounds.[40] First, because most of the sperm remain outside his wife's reproductive tract, they view the procedure as entailing a waste of semen. Procurement of sperm in IVF is said to violate the norm prohibiting the destruction of seed, particularly the ban on masturbation in all situations. Second, they oppose IVF because it diverges dramatically from the natural process of reproduction, unacceptably upsetting and unjustifiably tampering with the order of procreation, making it impossible to view a husband and wife as the genuine parents of the offspring. They view its acceptance as a legitimate form of reproduction as leading humanity on a slippery slope to moral degradation.

Questions about status also exist, specifically for a child who is genetically not Jewish. One Reform responsum considered the situation where sperm from a non-Jewish husband fertilizing an egg from a non-Jewish donor in vitro. The fertilized egg was then implanted in his Jewish wife. Because the fetus was part of its mother during the period of gestation, the responsum concluded that this child, born of a Jewish mother, is "potentially Jewish without any formal act of conversion"; however, the child would need to participate in positive acts of Jewish identification.[41]

Despite these objections and possible lineage questions, a broad consensus in support of IVF has emerged. As we have seen, most traditional authorities support AIH, while expressing a strong preference for other techniques to procure semen. They allow masturbation when required for some forms of artificial reproduction, including IVF. There is no emission of seed in vain, traditionalists generally conclude, because the husband intends to use his sperm to procreate.

The overwhelming majority of Jewish commentators do not condemn the technological transformation of the natural order of things. Marital reproductive procedures are desirable, but not absolute. As Reform Rabbi David Ellenson writes, "Reproductive interventions are no different in kind or degree from other interventions in the natural order."[42] These techniques, such as IVF, increase reproductive possibilities, enabling infertile couples to have children.

The medical risks to women or fetuses through the use of IVF procedures are not excessive or unacceptable. The Book of Deuteronomy (4:15) tells us to guard our lives carefully. The rabbinic tradition says that a person should not endanger one's life or expose oneself to excessive risk (Babylonian Talmud Shabbat 32a). The Babylonian Talmud (Hullin 10a) further tells us that something which is dangerous is to be avoided even more carefully than that which is ritually forbidden.

Techniques to retrieve eggs and transfer an egg or an embryo to the uterus are basically non-invasive and do not require use of a general anesthetic. The risks associated with the drugs that promote ovulation must, however, be evaluated on an individual basis.

From the perspective of Spiritual Judaism, the use of IVF is compassionate and responsible. It lessens a childless couple's anguish and emotional suffering. It enables an infertile heterosexual couple, a single man or woman, or a homosexual couple to experience many of the genetic, gestational, and social aspects of childbearing and parenting. The use of IVF accords with our responsibility to be active in partnership with God in improving and perfecting the world.

Looking to the future, IVF may possibly become quite common, at least among the wealthier; it may become the norm. As we will see in Chapter 8, IVF allows a choice to be made among an array of embryos outside the womb for implantation. A couple can choose from a range of different combinations of their genes (or those provided by sperm or egg donors). They can select an embryo that they view as not "deficient"; an embryo with some combination of genes that looks "good" and "healthy."

Single Motherhood, Artificial Insemination and In Vitro Fertilization

With the divorce rate exceeding 40% of all marriages, and people remaining single longer, at some point women no longer can depend on finding and marrying a Prince Charming for lifelong affection and support. For older, business or professional, financially stable and emotionally independent women, marriage may no longer be necessary if they cannot find a man they love, want to marry, and establish an emotional and economic partnership.

These women come to realize that the men who are available and their "jerk genes" may be more trouble than they are worth. They decide it's time to have children if they want them on their own. For them, the biological clock overrides social convention or religious tradition. Also, these women make enough money to support a child (or even children) without husbands; besides their parents, who may have no other prospect of grandchildren, may provide financial assistance.

Unmarried women make up an estimated 30% of the clients of sperm banks in the United States.[43] Single females use artificial insemination with donor sperm in vivo and in vitro. An unmarried woman's egg is fertilized by the sperm of a man she never met, but whom she selected for his genetic contributions to her child's potential success.

Jewish sages assumed it is best to have parents of two different genders because mothers and fathers perform distinctive roles. Apart from the traditional roles mothers and fathers play, children need two caring parents. Because everyone must deal with both genders in the adult world, having two sexes at home is an advantage. Two caring adults are better than one; together they likely will have more money and more time for their offspring.

Statistics and anecdotal evidence paint a grim picture of children born out-of-wedlock. They generally face significant disadvantages in educational achievement and personal development. They have greater problems with self-control and are often diagnosed as hyperactive; they are more likely to drop out of high school, get pregnant as teens, and abuse drugs.[44]

Children born to unmarried mothers are not inexorably inclined to be dysfunctional. Good single mothers can make a difference. However, children, on average, do worse with one parent especially when a single parent is isolated as the child's single caregiver. It is difficult to be a single mother, especially in the absence of grandparents, aunts and uncles, to fill the gap.

Because it takes time, money and energy to raise a child, a single woman contemplating motherhood, should ask herself: does she (alone or in combination with her parents) have sufficient funds to hire help? Or, if not, is there a strong support network of family and friends? If either exists, then despite the general statistics, children of single parents on average probably do as well as those with two parents.

The Jewish Tradition: Surrogacy

Despite a history of surrogacy in the Hebrew Bible, many text-oriented traditionalists condemn gestational and ovum surrogacy. Some liberal authorities raise objections to surrogacy but generally favor these arrangements.

Surrogacy in the Hebrew Bible

The authors of the Book of Genesis approved of third party-female intervention as a last resort in building a family. Three legendary matriarchs, Sarah, Rachel, and Leah, who had trouble conceiving and bearing children used surrogates, members of the extended family of their respective patriarchs to assure their husband's genetic continuity. The resulting offspring were treated as the patriarch's descendants with full inheritance rights.

Sarah said to Abraham (actually Abram, as he was then called), "God has kept me from having children. Why don't you sleep with my maid, Hagar? Perhaps she can have a child for me" (Genesis 16:2). Abraham agreed and Hagar gave birth to Ishmael. Unable to conceive, Rachel said to Jacob, "Here is my maid, Bilhah, have intercourse with her, so that she can have a child for me. In this way, I can become a mother through her" (Genesis 30:3). When Leah, Rachel's sister, was unable to continue to bear children, she gave her maid, Zilpah, to Jacob (Genesis 30:9). When the children were born to Bilhah and Zilpah, Rachel and Leah, respectively, adopted them and gave them their names (Genesis 30:6, 8, 11 and 13). Rachel's naming Dan and Naftali and Leah's naming Gad and Asher shaped their identities. Bilhah and Zilpah relinquished all rights to the children who became Rachel and Leah's sons.

Although these extraordinary efforts were justified as a last resort in assuring genetic continuity, these legends are not perfect parallels for us today. Marriage in ancient Israel was polygamous. As developed in Chapter 3, it also was socially acceptable for a man to conceive children with a woman, a concubine, in addition to his primary wife. A third party to a marriage was accepted as a surrogate to bear children. These surrogates were additional wives, part of an extended, polygamous family structure. In these arrangements, the child and the biological mother, who continued to play a role in the child's life, were part of the same household and family.

Surrogacy Condemned By Traditionalists

Many contemporary traditionalists condemn both gestational and ovum surrogacy. If a surrogate is married, these text-oriented traditionalists oppose the arrangement because it is viewed as adulterous, possibly leading to unintended incest if donated sperm or eggs are used, and raising doubts as to identity of the father if DI is utilized.[45] We have already considered each of these objections in discussing DI and egg donations, earlier in this chapter. An additional word or two is in order regarding the question of adultery.

A surrogate, whether married or unmarried, does not commit adultery because insemination takes place artificially; sexual intercourse does not occur. There is a lack of contact of genital organs and any intent to have an adulterous relationship. As one Reform responsum somewhat tentatively concludes: "We would therefore not consider the use of a married surrogate mother as adulterous, as the beginning of the process is akin to artificial insemination [and no sexual penetration would occur]. We would therefore hesitantly permit the use of a married surrogate mother in order to enable a couple to have children and await further clarification of medical and civil legal issues."[46]

Contemporary Quandaries

Apart from these technical objections, there are three other possible objections to surrogacy arrangements, namely, the exploitation of women, the payment of compensation for giving birth to a child, and destruction of the family.[47] As always, lineage issues must be considered.

Surrogacy Does Not Exploit Women

Surrogacy opponents maintain that exploitation inheres in surrogacy arrangements.[48] As Lord Jakobovits wrote, "To use another person as an 'incubator' and then take from her the child she carried and delivered for a fee is a revolting degradation of maternity and an affront to human dignity."[49]

Jewish law condemns exploitation. The rabbis (Babylonian Talmud Bava Meziah 59b) barred taking advantage of another's distress, weakness, or inexperience. With surrogacy, exploitation is seen in the abuse of poor women by wealthy couples, the frivolous avoidance of childbirth by fertile women, and the undue physical and psychological risks of pregnancy and relinquishment imposed on surrogates, thereby devaluing a woman's body.

Studies indicate that surrogate mothers are not poor; rather they are married, middle income women, predominantly motivated by altruism, a desire to help an infertile couple.[50] Although allowing women to spend more time to build a career or look for the right husband before having children, females have not used surrogates to avoid the risks of pregnancy. The intended parents are committed to their surrogate throughout the period of conception, gestation, and birth. They are concerned about, among other items, factors that could affect a fetus's health, the surrogate's physical and mental health, her emotional stability, her home life and more generally, her lifestyle.

A surrogate is exposed to the risks and pain of pregnancy and childbirth. A potential surrogate is free to assess these risks and weigh them against the benefits, financial and altruistic. A surrogate may gain fulfilment in enabling an infertile heterosexual couple (or single woman or lesbian couple, a single man or a gay couple) to be blessed with a child. Enabling someone who wants to have a child enhances family life, justifying and more than counterbalancing, for many prospective surrogates, the risks and pain.

Women are free to make their own decisions, whether or not to serve as a surrogate and terminate, preconception, her parental rights. As developed in Chapter 2, Spiritual Judaism wants people to be responsible and accountable for their choices. Surrogacy gives women the chance to help couples or single persons in need, thereby performing a good deed. Surrogates freely enter into these arrangements. They give their voluntary, informed consent before conception. There is no hurried decision-making or an unwanted pregnancy. There is no fundamental injury to their status as free persons. They know that they will be cut off from the child of their womb. Before entering into a surrogacy arrangement, they realize that they may later regret their decision. In advance of pregnancy, Spiritual Judaism recommends letting females determine what risks they want to face.

The typical agreement contains numerous protections for a surrogate. She is given the freedom to withdraw from the agreement prior to conception, guaranteed the right over her body during pregnancy, including the right to abort, and provided with payment of all expenses, medical, psychological counseling, attorney's fees, and living expenses. Her consent is informed. If the surrogate-to-be is married, most contracts require her to obtain her husband's consent because her pregnancy encumbers his marital rights.

Compensation For Surrogacy Is Not Improper

The arrangement typically provides for a surrogate to be compensated for her services. There is nothing wrong with this. Money is linked to pregnancy in that the ovum or gestational surrogate receives payment from the intended parents. For a surrogate, pregnancy involves lost time, some medical risks, and a degree of pain, warranting remuneration. Payment compensates a surrogate for the risks and hardships of pregnancy and helps assure that the gestational mother and the fetus receive proper care from the time of conception through delivery. Payment enhances the economic power of women.

Because more than twenty states have adopted surrogacy laws, parties contemplating a surrogacy arrangement should consult an attorney. Amid the vast array of statutes, there are four basic positions. First, one jurisdiction, the District of Columbia,[51] bans surrogacy contracts. Second, some states permit surrogacy arrangements, but paying compensation to a surrogate is illegal and may be subject to criminal sanctions.[52] Third, in other states make surrogacy contracts are not illegal, but are deemed void and unenforceable.[53] Fourth, some states find surrogacy contracts enforceable, provided certain regulatory measures are met.[54]

Surrogacy Will Not Lead to the Destruction of the Family

Surrogacy, it is asserted, may lead to disintegration of the family unit. Surrogacy opponents give great weight to procreation and family structure, resting on the preservation of the sanctity of marriage. By introducing a third party into a couple's reproductive process, they assert that surrogacy may undermine a married couple's relationship. According to one Reform Rabbi, Marc Gellman, "The sanctity of family life requires a single husband and single wife."[55]

However, most liberal commentators[56] and even some traditionalists[57] approve of surrogacy arrangements and have accepted surrogacy as a last resort for childless, married, heterosexual couples. As one Reform responsum states: "We would, therefore, have to treat the use of a surrogate mother as a new medical way of relieving the childlessness of a couple and enabling them to fulfill the *mitzvah* of procreation. It should cause us no more problems than modern adoptions which occur frequently....[58]

These progressive authorities and commentators support using the tools of medical science to promote procreation and overcome infertility which otherwise can impose an enormous emotional strain on marriages. They see surrogacy as bringing a solution to an infertile couple who want a child. There is no real threat to the well-being and integrity of the family receiving the child. Rather, the value of family life motivates an infertile couple to enter into a surrogacy arrangement to obtain a child who enhances their family. They derive great joy in gaining a child. They want to care for and nurture the child. Children born of surrogacy arrangements will usually be reared in circumstances "naturally" born children would envy.

From the perspective of Spiritual Judaism, a third party, a surrogate, who assists in procreation does not harm a marriage, as adultery does, but strengthens a heterosexual or homosexual couple's union. Surrogacy is compassionate. It enables infertile couples to receive a much wanted child, responding to the hurt infertility causes them, thereby enhancing and strengthening families. Surrogacy also makes it possible for single males, straight or gay, to impregnate artificially a woman who serves as an ovum surrogate and thus have families. Surrogacy thus facilitates family life by enabling married heterosexual couples (including postmenopausal women), single males (or females who lack the ability to carry a fetus to birth), and gay couples to have children. It offers and extends the joys of parenthood.

Finally, it is important to look at a surrogacy arrangement from the child's standpoint. The intended parents accept (or the single-parent accepts) responsibility for the child from the time of conception, thereby protecting a child's interest in entering a secure home, where he or she is wanted, loved, cared for, and nurtured. Although encountering some separation anxiety from his or her gestational mother, "psychic" bonds perhaps, a child born of a surrogacy arrangement is not reduced to a commodity.

Jewish Status Questions

The question "Is the child Jewish?" has to receive special consideration in the case of surrogate mothering. As discussed in Chapter 5, a child's religious status for traditionalists, including the Conservative Movement (but not the Reform and Reconstructionist Movements), derives from his or her gestational mother at the time of birth or through formal conversion to Judaism. The religious status of the birth mother determines Jewish identity in nonsurrogacy cases of DI and IVF, not involving egg donors.

Although things get more difficult with surrogacy or egg donation, a consensus has developed that the birth mother is the child's mother. Traditionalists and the Conservative Movement look to opening the womb as determinative.[59] Thus, if the gestational mother, who provides the womb, is Jewish, the child is Jewish regardless of the source of the egg used, even if non-Jewish or of uncertain religious status. Conversely, if the gestational mother is not Jewish (or her religious identity cannot be determined), the child is not Jewish and gains Jewish status only through conversion.

If the ovum surrogate (the genetic mother and the birth mother) is Jewish, the child is Jewish. Conversely, for traditionalists and the Conservative Movement, if the ovum surrogate is non-Jewish, the child is non-Jewish. After the intended parents receive the child, they must proceed to convert the newborn.

A Surrogate's Perspective

Prospective surrogates must be aware of several points. From a Jewish law perspective, Jewish ovum surrogates encounter the objections of traditionalists previously discussed in connection with egg donations, especially the possibility of incestuous marriages by the surrogacy offspring.

Despite the risks and physical dangers surrogates face, liberal commentators generally view with favor women serving as surrogates. A Jewish woman may serve as a gestational surrogate, liberal authorities have concluded, if she primarily intends to assist an infertile couple to have a child.[60] Furthermore, acting as a surrogate, whether gestational or ovum, represents an act of lovingkindness, so important to Spiritual Judaism. In deciding whether to serve as a surrogate, a woman must consider her medical risk factors, including her medical history and age.

With certain precautions, a surrogate's other children likely will not feel threatened by her contractual pregnancy. If a surrogate's children are told that the newborn will be the contracting couple's child, they will realize that they will not be relinquished, sold or abandoned. If properly explained, the surrogate's other children will see that a newborn is given away, not taken away, even though they may feel something emotionally that they cannot articulate.

The parties to a surrogacy arrangement should, if at all possible, obtain the consent of the surrogate's husband. Any husband of a pregnant woman is very involved in her experience and is required to do a lot for her. Even if he consents, there is a measure of emotion on his part, watching a pregnancy develop and knowing it will not be his child.

Spiritual Judaism welcomes the use of assisted reproductive techniques discussed in this chapter. It favors these modern means of procreation, although they depart from the usual method of producing a child through intercourse and they separate reproduction from sex. They allow infertile couples, single women and men, gay or lesbian couples to experience the blessing of children, the social aspects of parenting, and sometimes its genetic and biological aspects. These technologies offer the potential for bringing a new child into the world, providing life to a child, adding joy to a family (or a single parent), and contributing strength to the Jewish community. These techniques are also compassionate. They alleviate an infertile couple's (or single parent's) pain and suffering -- their shared distress. Because our desire for offspring is deeply human, helping an infertile heterosexual couple, a single individual, or a homosexual couple realize their desire is a compassionate act. These new technologies support the diverse family types springing up in America, including gay and lesbian families and single parents.

Assisted reproduction techniques, especially in vitro fertilization, serve as a stepping stone to reproductive possibilities far beyond treating infertile couples or offering single women (or men) or gay or lesbian couples the chance to have a child. IVF marks the start of human beings assuming power to control their procreative and evolutionary destinies, topics considered in Chapters 7 and 8.

Notes

1. U.S. Department of Health and Human Services, Centers for Disease Control and Prevention/National Center for Health Statistics, *Fertility, Family Planning, and Women's Health: New Data From the 1995 National Survey of Family Growth*, Series 23, No. 19 (Washington: Government Printing Office, 1997), Table 51. Number of Currently Married Women 15-44 Years of Age and Percentage Distribution by Infertility Status, According to Selected Characteristics: United States, 1995, 61; Congress of the United States, Office of Technology Assessment, *Infertility: Medical and Social Choices* (Washington: Government Printing Office, 1988), 3, 51.
2. *Id.*, 7; *The Women's Health Data Book: A Profile of Women's Health in the United States*, ed. Jacqueline A. Horton (Washington, D.C.: Jacobs Institute of Women's Health/Elsevier, 1992), 25.
3. Office of Technology Assessment, *Infertility: Medical and Social Choices*, 9.
4. Barbara Vobejda, "Egg Donation: A Growing Business," Washington Post, March 7, 1999, A1, A7, Gino Kolata, "$50,000 Offered to Tall, Smart Egg Donor," *New York Times*, March 3, 1999, A10.
5. I have drawn on David S. Shapiro, "Be Fruitful and Multiply" in *Jewish Bioethics*, eds. Fred Rosner and J. David Bleich (New York: Hebrew Publishing Co., 1979).
6. *Id.*, 64-70.
7. Rabbi Shlomo Ganzfried, *Kitzur Shulchon Oruch*, Volume II, trans. Rabbi Eliyah Touger (New York: Moznaim, 1991), Chapter 145:2, The Laws of Marriage, 605.
8. *The Code of Maimonides*, Book Four, The Book of Women, trans. Isaac Klein (New Haven: Yale University Press, 1972), Treatise I, Law Concerning Marriage, Chapter 15:16, 97.
9. Immanuel Jakobovits, *Jewish Medical Ethics: A Comparative and Historical Study of the Jewish Religious Attitude to Medicine and Its Practice* (New York: Bloch, 1975), 264; J. David Bleich, *Judaism and Healing: Halakhic Perspectives* (New York: KTAV, 1981), 82-83; J. David Bleich, *Bioethical Dilemmas: A Jewish Perspective* (Hoboken, NJ: KTAV, 1998) 220, fn. 4 and 5, at 233-234; Fred Rosner, "Artificial Insemination in Jewish Law" in *Jewish Bioethics*, eds. Fred Rosner and J. David Bleich (New York: Hebrew Publishing Co., 1979), 112,

fn. 45, at 117 (citing Rabbi Moshe Feinstein); Fred Rosner, M.D., *Modern Medicine and Jewish Ethics*, second ed. (Hoboken, NJ: KTAV, 1991), 94. However, Rabbi Eliezer Waldenberg frowns on AIH, asserting it is permissible only in extreme situations. Rosner, "Artificial Insemination," in *Jewish Bioethics*, 112; Rosner, *Modern Medicine and Jewish Ethics*, 94.

10. Moshe Zemer, "A Rabbinic Ban On Sperm Donations," in *The Fetus and Fertility: Essays and Responsa*, ed. Walter Jacob and Moshe Zemer (Pittsburgh: Rodef Shalom Press, 1995), 164.

11. Daniel J. Lasker, "Kabbalah, Halakhah, and Modern Medicine," Modern Judaism 8:1 (February 1988):1-14, at 3-4. Reform Rabbi Solomon Freehof noted the Kabbalistic belief that certain lives created by a wasteful emission of sperm are distorted lives, evil spirits. Responsum No. 47, "Insemination With Mixed Seed," in Solomon B. Freehof, *New Reform Responsa* (Cincinnati: Hebrew Union College Press, 1980), 207.

12. Rosner, *Modern Medicine and Jewish Ethics*, 95; Bleich, *Bioethical Dilemmas*, 221-224.

13. Yoel Jakobovits, "Male Infertility: Halakhic Issues in Investigation and Management," *Tradition* 27:2 (Winter 1993):4-21, at 8-13, reprinted in *Jewish Law and The New Reproductive Technologies*, ed. Emanuel Feldman and Joel B. Wolowelsky (Hoboken, NY: KTAV, 1997); Rosner, *Modern Medicine and Jewish Ethics*, 95-96; Rosner, "Artificial Insemination," in *Jewish Bioethics*, 114-115; Bleich, *Judaism and Healing*, 83; Bleich, *Bioethical Dilemmas*, 249-250; J. David Bleich, "Test-Tube Babies," in *Jewish Bioethics*, 83; J. David Bleich, "Surrogate Motherhood," *Tradition* 32:2 (Winter 1998): 146-163, at 154.

14. Central Conference of American Rabbis (CCAR), Responsum No. 153. "Masturbation (1979)," in *American Reform Responsa: Collected Responsa of CCAR 1889-1983*, ed. Walter Jacob (New York: CCAR, 1983), 480.

15. Elliot N. Dorff, *Matters of Life and Death: A Jewish Approach to Modern Medical Ethics* (Philadelphia: Jewish Publication Society, 1998), 53. See also Elliot N. Dorff, "Artificial Insemination, Egg Donation and Adoption," *Conservative Judaism* 49:1 (Fall 1996): 3-60, at 55.

16. Rabbi Alfred S. Cohen, "Artificial Insemination," *Journal of Halacha and Contemporary Society 13* (Spring 1987):43-59 provides a helpful summary of controversy two among leading rabbinic decisors, Rabbi Moshe Feinstein and Rabbi Yaakov Breisch.

17. Bleich, *Bioethical Dilemmas*, 208-209; Bleich, "Surrogate Motherhood," 152-153; Rosner, "Artificial Insemination," in *Jewish Bioethics*, 110, fn. 22, at 116; Rosner, *Modern Medicine and Jewish Ethics*, 90-91, fn. 19, at 98.

18. Bleich, *Judaism and Healing*, 81, fn. 1 and 2, at 84 (citing Rabbi Moshe Feinstein requiring genital contact for a sexual act to be adulterous; however, Rabbi Eliezer Waldenberg did not require sexual contact as long as one man's semen enters another's wife); Bleich, "Surrogate Motherhood," 150; Bleich, *Bioethical Dilemmas*, 244, fn. 12, at 263; Rosner, "Artificial Insemination," in

Jewish Bioethics, 109, fn. 15, at 116. Rabbi Bleich opposes DI, viewing it as quasi-adulterous in nature, as a prohibited form of procreation that is contrary to the spirit of the law. Bleich, *Bioethical Dilemmas*, 246; Bleich, "Surrogate Motherhood," 151.

19. Noam J. Zohar, *Alternatives in Jewish Bioethics* (Albany: State University of New York, 1997), 71-78 (citing Rabbi Eliezer Waldenberg and Rabbi Moshe Feinstein); Noam Zohar, "Artificial Insemination and Surrogate Motherhood," S'vara: A Journal of Philosophy and Judaism 2:1 (Winter 1991):13-19, at 15-17.

20. Dorff, *Matters of Life and Death*, 68-69; Dorff, "Artificial Insemination," 19-20, 55.

21. Rosner, "Artificial Insemination," in *Jewish Bioethics*, 110; Rosner, *Modern Medicine and Jewish Bioethics*, 91; Bleich, *Judaism and Healing*, 82.

22. *The Code of Maimonides*, Book Five, The Book of Holiness, trans. Louis I. Rabinowitz and Philip Grossman (New Haven: Yale University Press, 1965), Treatise I, Laws Concerning Forbidden Intercourse, Chapter 2: 1-5, 7-9, at 3-4, 17-19; Fred Rosner, M.D., *Sex Ethics in the Writings of Moses Maimonides* (New York: Bloch, 1974), 96.

23. CCAR Responsum No. 197. "Child Born Through Artificial Insemination (1986)" in Walter Jacob, *Contemporary American Reform Responsa* (New York: CCAR, 1987), 292.

24. CCAR Responsum No. 157. "Artificial Insemination (1952)" in *American Reform Responsa*, 501. A contemporaneous Reform responsum, CCAR Responsum No. 158. "Artificial Insemination (1952)", *id.*, 504 offered a more guarded conclusion: "While indications strongly point to a negative answer [to artificial insemination] (particularly if the seed of a stranger is to be used), other aspects of Judaism must be explored as well, in order to arrive at a conclusion reflecting Judaism at its best." More recently, a Reform commentator summarized the complex moral and Jewish concerns surrounding this technology. David Ellenson, "Artificial Fertilization (*Hafrayyah Melakhotit*) and Procreative Autonomy: *In Light of Two Contemporary Israeli Responsa*" in *The Fetus and Fertility*, 35.

25. Jakobovits, *Jewish Medical Ethics*, 264. See also Rosner, *Modern Medicine and Jewish Ethics*, 91 (citing Rabbi Eliezer Yehuda Waldenberg).

26. Lasker, "Kabbalah, Halakhah, and Modern Medicine," 7-9.

27. *Uniform Laws Annotated: Matrimonial, Family and Health Laws*, vol. 9B, at 287.

28. These state statutes include Florida Statutes §742.14; North Dakota Centennial Code §14-18-04, Oklahoma Statutes Title 10 §555; Texas Family Code §151.102; and Virginia Code §20-158(A).

29. Sheryl Gay Stolberg, "Quandary on Donor Eggs: What to Tell the Children," *New York Times*, January 18, 1998, Section 1, at 1.

30. Jakobovits, *Jewish Medical Ethics*, 248-249.

31. Dorff, *Matters of Life and Death*, 81, 95-96; Dorff, "Artificial Insemination,"

40-41, 54.

32. CCAR Responsum No. 19. "*In Vitro* Fertilization with Cousin's *Ova* (1986)," in *Contemporary American Reform Responsa*, 32.

33. Susan Golombok et al, "Families Created by the New Reproductive Technologies: Quality of Parenting and Social and Emotional Development of the Children," *Child Development* 66:2 (April 1995): 285-298; Susan Golombok and Clare Murray, "Social versus Biological Parenting: Family Functioning and Socioemotional Development of Children Conceived by Egg or Sperm Donation," *Journal of Child Psychology* 40:4 (May 1999): 519-527.

34. J. David Bleich, "In Vitro Fertilization: Questions of Material Identity and Conversion," *Tradition* 25:4 (Summer 1991): 82-102, at 94-95.

35. *Id.*, at 94.

36. Dorff, "Artificial Insemination," 44.

37. Bleich, *Judaism and Healing*, 91; Bleich, "Test-Tube Babies," in *Jewish Bioethics*, 84.

38. Aaron L. Mackler, "An Expanded Partnership with God? In Vitro Fertilization in Jewish Ethics," Journal of Religious Ethics 25:2 (Fall 1997): 277-304, at 297, a responsum approved by the Conservative Movement's Committee on Jewish Law and Standards. Dorff, *Matters of Life and Death*, 55.

39. CCAR Responsum No. 18. "Test Tube Baby (1978)," in *Contemporary American Reform Responsa*, 30.

40. Eli D. Clark and Ze'ev Silverman, "Surrogate Motherhood in the Case of High-Risk Pregnancy," Journal of Halacha and Contemporary Society 38 (Fall 1999): 5-38, at 21; Mackler, "An Expanded Partnership with God?", at 283 (citing Rabbi Eliezer Yehudah Waldenberg); David Ellenson, "Artificial Fertilization" in *The Fetus and Fertility*, 21-25.

41. CCAR Responsum No. 41. "Status of a Child Genetically Not Jewish (1984)," in *Contemporary American Reform Responsa*, 71.

42. Ellenson, "Artificial Fertilization," in *The Fetus and Fertility*, 32.

Jonathan Eig, "An American Family," *Offspring* 1:4 (December/January 2001): 92-101, at 94.

43. Jonathan Eig, "An American Family," *Offspring* 1:4 (December/January 2001): 92-101, at 94.

44. James Q. Wilson, "Human Remedies for Social Disorders," The Public Interest Number 131 (Spring 1998): 25-35; Lee Smith, "The New Wave of Illegitimacy," *Fortune* 129:8 (April 18, 1994): 81-94, at 82 (citing data on thousands of children collected by U.S. Department of Health and Human Services).

45. Bleich, *Bioethical Dilemmas*, 246-247.

46. CCAR Responsum No. 159, "Surrogate Mother (1982)," in *American Reform Responsa*, 507.

47. I have drawn on Elie Spitz, "Through Her I Too Shall Bear a Child: Birth Surrogates in Jewish Law," Journal of Religious Ethics 24:1 (Spring 1996): 65-

97, at 79-87. This responsum was adopted by the Conservative Movement's Committee on Jewish Law and Standards. Dorff, *Matters of Life and Death*, 62. Lori B. Andrews, "Surrogate Motherhood: The Challenge for Feminists," Law, Medicine & Health Care 16:1-2 (Spring-Summer 1988): 72-80, provides a helpful overview and rebuttal of these assertions.

48. Marc Gellman, "The Ethics of Surrogate Motherhood," *Sh'ma* 17:334 (May 15, 1987): 105-107, at 106.

49. Jakobovits, *Jewish Medical Ethics*, 265.

50. Dorff, *Matters of Life and Death*, 63, fn. 73, at 346; Spitz, "Through Her I Too Shall Bear a Child," 67, 81.

51. District of Columbia Code §16-402.

52. These state statutes include Kentucky Revised Statutes §199.590(4) and Utah Code §76-7-204. The Michigan statute provides for the criminalization of gestational surrogacy contracts involving compensation. Michigan Compiled Laws §722.859. Seven states, Florida, Michigan, Nevada, New Hampshire, New York, Virginia, and Washington, ban payments to surrogates, but provide a wide variety of exceptions allowing the payment of a surrogate's expenses. Florida Statutes §742.15(4); Michigan Compiled Laws §722.853; Nevada Revised Statutes §126.045(3); New Hampshire Revised Statutes §168.B:25(V); New York Domestic Relations Law §123(1); Virginia Code §§20-156, 20-160(B)(4)-(5); Washington Revised Code §§26.26.210(1),.230.

53. The state laws include Indiana Code §31-20-1-2; Louisiana Revised Statutes §9:2713; Michigan Compiled Laws §722.855; Nebraska Revised Statutes §25-21, 200; North Dakota Century Code §14-18-05.

54. Florida Statutes §742.15; Nevada Revised Statutes §126.045; New Hampshire Revised Statutes §168-B:16-B:28; Virginia Code §20-156 to -165. New Hampshire and Virginia require, among other provisions, advance judicial approval of a surrogacy agreement.

55. Gellman, "The Ethics of Surrogate Motherhood," 107.

56. Spitz, "Through Her I Too Shall Bear a Child," a responsum adopted by the Conservative Movement's Committee on Jewish Law and Standards; Dorff, *Matters of Life and Death*, 62; CCAR Responsum No. 159. "Surrogate Mother (1982)," in *American Reform Responsa*, 506-507. However, Aaron L. Mackler, "Comment on 'Through Her I Too Shall Bear A Child': Birth Surrogates in Jewish Law," *Journal of Religious Ethics* 27:2 (Summer 1999): 361-366, expresses a number of reservations regarding surrogacy arrangements, including, first, what if the intended parents refuse custody of a child of an undesired gender or if he/she has a birth defect or an unexpected disability and, second, will other children of a surrogate be adversely affected? Mackler focuses on the potential for psychological harm to surrogate's children, who see their mother go through pregnancy and give birth to a child who is given over to others. He sees the imposition of a foreseeable harm on some children, without their consent. Mackler would only permit surrogacy arrangements in exceptional cases

involving immediate relatives.

57. Rosner, *Modern Medicine and Jewish Ethics*, 114; Clark and Silverman, "Surrogate Motherhood in the Case of High-Risk Pregnancy," 38.

58. CCAR Responsum No. 159. "Surrogate Mother (1982)" in *American Reform Responsa*, 506.

59. Dorff, *Matters of Life and Death*, 101; Bleich, "In Vitro Fertilization," at 82, 84, 95; Mackler, "An Expanded Partnership with God?", 295-297. In "In Vitro Fertilization," 86-88, 94-95, Rabbi Bleich propounds a possible theory of dual maternal relations. Gestational surrogacy offspring may have two mothers with the genetic mother and the birth mother having maternal status, thus recognizing the contributions of both to a child's birth. Bleich asserts that the host mother is legally recognized as the mother because of her role in nurturing and sustaining the fetus. The genetic mother should also be considered a mother because of her significant role in the production of the child. See also J. David Bleich, *Contemporary Halakhic Problems* (New York: KTAV, 1997), 108; Bleich, *Judaism and Healing*, 94. Perhaps because granting maternal status to two women with respect to one child creates the possibility of conflict, Bleich concludes that the genetic relationship is doubtful. Bleich, "In Vitro Fertilization," 95. One commentator would allow a genetic mother to give her religious identity to the child. Because of the genetic mother's close feelings for the child and viewing him or her as an extension of the Jewish genetic-intended mother, Spitz, "Through Her I Too Shall Bear a Child," 90-91, would define the newborn as a Jew.

60. Dorff, "Artificial Insemination," 56.

Chapter 7

Human Cloning: The Advent of Asexual Reproduction

In the twenty-first century, scientists will have perfected human cloning, copying the DNA (the chemical code programming each of the body's cells) from one human to create another asexually. Perhaps much sooner than any of us realize, we will see nearly genetically identical replicas of humans.[1]

In this chapter, I want to discuss the views of the Jewish tradition on this matter. Let us, therefore, suppose that the process will be developed. Let us further assume that it will be safe and effective.[2] The new living beings this generated, as children and adults, will be physically healthy. They will not be more at risk of being abnormal at birth. They will not age faster than normal or succumb prematurely to diseases of old age. They will not suffer greater than normal psychological problems as a result of being a clone. All this will constitute my hypothesis for bringing out the various ethical arguments.

Human cloning will offer therapeutic benefits. A clone's cells could provide a source for tissue transplants and other medical needs. The cells taken from a clone would possess nearly identical immunities, thereby eliminating, or at least drastically reducing, the risk of rejection, one of the major problems currently arising from tissue transplants.

Human cloning will also provide a number of beneficent reproductive scenarios. It will help incurably infertile couples, including women without viable eggs or men having no viable sperm, unable to be aided by, or dissatisfied with, other modern reproductive techniques, discussed in Chapter 6, or those facing a high risk of conceiving children with one or more severe genetic disorders, not amenable to genetic screening and treatment. For instance, prospective parents who are both Tay-Sachs carriers could avoid the risk of a doomed child. A cloned child, as a genetic copy of one parent, would be a carrier of this disease but would not suffer from the fatal malady. Even if only one spouse has a serious genetic disease or weakness, cloning the other spouse permits procreation without fear of transmitting a bad gene or a serious hereditary disease. (However, as we will see in Chapter 8, the selection of embryos or genetic engineering, rather than cloning, may more easily prevent the transmission of deadly or defective genes from one or both parents.) Cloning could also be used to reproduce someone with special meaning, such as an infertile widow who wants to clone her only dying child.

Numerous other possible procreative scenarios exist, including a hearty, if not narcissistic, ego intent on reproducing itself without a partner's genes, homosexual couples or single individuals seeking a child, those desirous of a good genetic legacy, extraordinary athletic or artistic talent, genius or character, or reducing the possibility of genetic disease by the selective choice of ancestor-genes, or aging individuals striving for immortality who would hire surrogates to bear their genetic offspring. However, it seems unlikely that cloning will be used to provide slaves for their creators, by dictators to perpetuate themselves, or to produce a warrior class to serve the purposes of the state or a demented ruler.

Rather than focusing on our most far fetched fears, this chapter, after briefly outlining the cloning process, develops the permissive approach of the Jewish tradition to technology, in general, and to human cloning, in particular. It then focuses on examining what really concerns the Jewish tradition, preservation of the traditional family, lineage, and the possibility of adultery. The chapter concludes by critically examining and refuting more general objections to human cloning.

The Cloning Process

In February, 1997, Dr. Ian Wilmut and his colleagues published a paper describing the birth in July, 1996 of a sheep, named Dolly, by cloning from a mature, nonreproductive cell.[3] Basically, the scientists injected this cell's nucleus, containing a full set of chromosomes, into an unfertilized host female's egg whose nucleus had been extracted and destroyed. The cell to be cloned was taken from the mammary gland of an adult sheep and sustained in a solution, but starved of nutrients in order to stop its development. In the cloning process, the selected cell was not a gamete, reproductive cell; rather, it was a body (somatic) cell that had undergone differentiation into a specialized cell that forms part of a tissue or an organ. This cell is called "mature" because it had reached its final cellular destination, turning off other regions of its DNA.

The scientists, who produced Dolly, knew that host egg's specialized interior, its cytoplasm, contained mitochondrial DNA, the energy factories of cells, able to activate all the potential genes contained in the nucleus. Thus, the DNA of the selected adult cell would develop into an entire organism.

The scientists used electric shocks to fuse the donor's nucleus with the host egg from which the nucleus had been removed so that it behaved like a newly fertilized egg would with genetic material from a male and a female. After fusion, the genetic information was reset and made young again. The cellular division cycle began leading to the development of an embryo. After maintaining the preembryo of about one hundred cells in a laboratory for six days, the scientists implanted it into the uterus of the gestational surrogate mother ewe. The implanted nucleus directed the formation of the specialized cells. The surrogate mother nurtured the fetus, carried it to term as in like a normal pregnancy, and gave birth to a healthy sheep, Dolly.

For humans, the mature cell's nucleus could come from a woman who has viable eggs and bears the child, from her spouse, or from a total stranger, male or female. Thus, a woman could produce a clone using her own genetic material and one of her own eggs. In this situation, the clone would have the same DNA make up as the donor and would be an exact genetic replica of her mother. If the donor of the genetic material and the egg donor were not the same female, the child, genetically-speaking, would be the mature cell nucleus donor's nearly identical twin, differing slightly because of the small amount of DNA that creates certain activating proteins in the egg. Obviously, a male can only serve as a nucleic donor, not an egg donor or a womb provider.

American Legal Background

The announcement of Dolly's birth in February, 1997 triggered a flurry of political activity in the United States. Almost immediately after the announcement, President William Jefferson Clinton ordered the National

Bioethics Advisory Commission (NBAC), an eighteen member panel, evenly balanced between scientists and nonscientists previously appointed by the President, to conduct a thorough review of the legal and ethical issues associated with human cloning and make recommendations for possible Federal actions to prevent its abuse. The President instructed the Commission to report to him in ninety days.[4]

One week later, President Clinton, taking decisive action, extended the existing federal funding ban on human embryo research, discussed in Chapter 9, to include research aimed at cloning humans. He barred all Federal agencies from supporting, funding, or undertaking such activity. He also urged a voluntary moratorium by all privately and publicly funded research on human cloning until the NBAC "and our entire Nation have had a real chance to understand and debate the profound ethical implications of the latest advances."[5]

The Report of the NBAC, delivered a little more than three months later, recommended that the federal government continue its moratorium on the use of federal funding in support of any attempt to clone a human being and that it request all firms, individuals, and professional societies in the private, nonprofit, and nonfederally funded sectors to comply voluntarily with the spirit of the federal moratorium. The Report also recommended that the federal government enact a legislative ban on any attempt, in a research or clinical setting, to create a human being by cloning.[6]

Following these recommendations, President Clinton proposed legislation prohibiting human cloning experiments in the United States by anyone in the private or public sectors involving the implantation of an embryo in a women's womb with the intent of making a baby. The bill contained a five year sunset provision after which the issue could be revisited.[7]

Although the moratorium banning federal funding on research aimed at cloning humans remains in place, the proposed legislation was not enacted on the federal level.[8] At present (September 2001), human cloning remains legal in the United States, except in a few states, including California, Louisiana, Michigan and Rhode Island.[9] For example, California adopted a temporary ban on cloning, prohibiting the creation of (or the attempting to create) children by transferring the nucleus of any human cell to a denucleated human egg cell for the purpose initiating a pregnancy.

Clinical researchers must, however, deal with the U.S. Food and Drug

Administration (FDA). The FDA has asserted its jurisdiction over the use of technology to clone a human being. October 1993, the FDA published a notice clarifying the application of FDA statutory authorities to human cloning.[10] The notice stated that somatic cell therapy products were biological products under the Public Health Service Act as well as drugs under the Federal Food, Drug, and Cosmetic Act and thus were subject to investigational new drug (IND) application requirements. After proposing a more comprehensive regulatory approach in March 1997, the FDA, in January 2001, issued a final rule establishing its regulatory approach for human cells, tissue, cellular and tissue-based products.[11] In brief, clinical researchers using technology to clone a human being, the FDA continued to assert, is subject to FDA regulation. Before such research may begin, a researcher must submit an IND request to the FDA, which the FDA will review to determine if the research can proceed.

Researchers and clinicians must also consider current medical ethical guidelines. Although the use of reproductive human cloning by medical professionals does not currently "meet standards of ethical acceptability," privately funded experimentation will occur in the United States, unless barred by federal legislation, particularly research into therapeutic human cloning not involving "transferring embryos to the uterus, provided that ethical procedures for conducting research are followed."[12] These ethical standards require investigators using gametes for research to respect the principle of informed consent.

The Permissive Approach of the Jewish Tradition to Technological Advances

In assessing technological advances, the Jewish tradition moves from the premise that any form of conduct not explicitly forbidden in the Torah or by the letter or the spirit of Jewish law is permissible.[13] There seems to be no direct halakhic prohibition of the cloning of humans or asexual procreation without the pairing, in some way, of a married man and woman. The Babylonian Talmud (Sanhedrin 65b) even relates that Rabbi Hanina and Rabbi Oshia met every Friday evening to create a calf for their Sabbath meal. Thus, the sages regarded asexual reproduction, at least with respect to animals, as acceptable and morally innocuous.

In a helpful partnership with God, humans are permitted, and in some sense are obligated, to build and develop the world for its betterment. We are God's active partners in bringing the ongoing creative process to completion. For traditionalists, the Eternal granted humans dominion over the world. According to the author of the Book of Genesis (1:28), the Holy One exhorted humans to fill the earth with their descendants who will bring the planet and all its inhabitants under their control. Unless specifically proscribed, we are to master and harness the earth and all of nature for humanity's benefit. Whatever we can use for human enhancement does not trespass on the Eternal. As the Psalmist reminds us: "Heaven belongs to the Holy One alone, but the Eternal gave the earth to humans" (Psalm 115:16).

As noted earlier in this chapter, cloning offers potentially great medical benefits including creating tissue-matched donors. We read in the Book of Exodus (21:19) that Jews are permitted, if not obligated, to use all available technology to heal the sick and attempt to restore a sufferer's lost health (Babylonian Talmud Baba Kamma 85a). In so doing, we are not tampering with the Divine order of creation.

By helping solve infertility problems, cloning will enable humans be fruitful and multiply; however, the process will contribute, at least to a limited degree, to already existing population problems. In other procreative scenarios, cloning creates neither a different human life form nor a new species. God bestows a unique soul in each cloned fetus. For Spiritual Judaism, a clone will thus be a child of God and be loved by the Eternal.

In using cloning technology, humans will not usurp the Divine prerogative anymore than with assisted reproduction techniques. No threat exists to the Divine nature of creation. We are neither playing nor replacing God. We are creating something out of something. Because the Eternal is in each of us, through cloning, we participate in the ongoing process of creation, hopefully in a prudent, sensible and responsible manner.

A Jewish Concern: Human Cloning and the Changing of The Nature of The Family Unit

Even if the Jewish tradition does not yet explicitly prohibit cloning, some may argue that it should be prohibited because use of this technology could lead to more serious violations of Jewish law. Opponents of human cloning maintain it poses a significant threat to the nuclear family structure and to procreation through married male-female sexual intercourse. Underpinning these concerns rests a view of the family and the nature of procreation. The Jewish tradition wants to safeguard the family unit, consisting of a married husband and wife, who produce offspring in a "natural" manner.

In the Babylonian Talmud (Kiddushin 30b and Niddah 31a), the sages remind us that there are three partners in creating a human being, the father, the mother, and God. Cloning may be seen as undermining this tripartite partnership. However, the rabbis simply noted a preference for what they saw as the natural way of the world. Although most humans will be formed as the product of a married couple, consisting two sexually different parents, who each contribute an egg or sperm to the process, we are not obligated to procreate only in this way.

In attempting to bar certain types of cloning, traditionalists may look to Genesis Rabbah 1:26 where we read: "There is no possibility for a man without a woman, or for a woman without a man." Thus, they may strive to prohibit cloning a child, at least using an egg and a cell containing genetic material from one or two women. In other words, a clone created from a female's egg and a male's genetic cell seems permissible, especially for married heterosexual couples, where the wife-mother carries the fetus to birth and the couple raises the child. Cloning may also raise fewer problems for traditionalists in contrast to two assisted reproduction techniques, AIH or DI, requiring the extraction of sperm, among other issues.

Because cloning separates reproduction from sex, it offers the possibility of a single woman or a lesbian couple giving birth to a female clone, thus defeating the notion of bearing children within a traditional, nuclear family structure. Or, a single man or one member of a gay couple could act in association with at least one woman (at least until technology comes up with an artificial womb), who will provide an egg and serve as the gestational surrogate for a male clone.

Cloning will thus empower women, who can gestate children with their own genes; men cannot. If they want a female-clone-child, women no longer need male participation in the reproductive process. Thus, cloning will benefit women by reducing their dependence on men's sperm. For a male to reproduce himself, he needs a willing woman (or women) to provide an egg and gestate the embryo with his genes. The egg provider also will contribute her mitochondrial DNA to a male clone-child.

Because cloning enables women to reproduce without men, asexual lesbian reproduction raises the prospect, albeit far into the future, of the development of an ongoing female-only community unencumbered by men.[14] In addition to upsetting the overall sex ratio, the prospect of women-only families raises the specter, however farfetched, of men becoming an endangered species.

Because sex is a staple of life and the overwhelming percentage of men and women are heterosexual, it seems highly unlikely that conceiving children the traditional way will go out of fashion. Married heterosexual couples, who are able, will want to produce children through intercourse or assisted sexual reproduction techniques. Cloning, in and of itself, will not produce a major change in the basic family unit. Spiritual Judaism urges that the prospect of non-nuclear families, particularly, single persons or gay or lesbian individuals (or couples), having clone children ought not to lead to prohibiting asexual reproduction.

Just as with the assisted reproduction techniques discussed in Chapter 6, the genie is already out of the bottle today with respect to the non-nuclear family. The past several decades have witnessed a remarkable transformation of the "family." The twenty-first century will continue to witness the flourishing of diverse of family structures, including non-traditional families consisting of no dad or two dads, perhaps with no one form dominating the mix. Single females may clone themselves to create offspring without turning to the unknown genetic material provided by anonymous sperm banks.

Although encouraging the formation of traditional families and helping them remain intact, Spiritual Judaism urges that we respect the value of personal choice and one's sexual orientation. As child rearers, gay or lesbian couples (or for that matter gay or lesbian single parents) will likely do no worse than one or two heterosexual parents. Based on research findings, one expert concluded, "[C]hildren of lesbian and gay parents develop as successfully as do children of heterosexual parents."[15]

Overcoming its fear of a diversity of family structures, Judaism needs to focus on facilitating the development of a stable and loving parental environment, whether by heterosexual couples, single parents, or homosexual couples. As the involved professionals, including rabbis, physicians, genetic counselors, clergy, attorneys, and social workers, advise and counsel couples (and individuals) regarding the social and psychological challenges of cloning and gain the potential parents' informed consent, those providing expert guidance ought to emphasize the importance of the child-rearing responsibilities, so important to

Spiritual Judaism. The parents (or parent) of a clone must want to love, care for, and nurture the clone-child as well as provide him or her with structure and discipline.

In the future assortment of family structures, some may still opt for procreation by the reproductive lottery, a randomly combined set of genes, over cloning that would produce more predictable offspring. The mix of genes, even from autonomous egg or sperm donors, often produces fascinating combinations, some of which may, however, prove quite harmful to the offspring. In Chapter 8, we will deal with new prenatal techniques, namely, genetic screening and engineering, designed to deal with these defects. Viewing a child as too important to leave to the chance meeting of sperm and egg, others may choose to procreate by cloning.

Lineage: Another Jewish Concern

Even if the Jewish tradition concludes that human cloning is permissible, lineage and parentage, especially knowing one's mother and father, remain important issues. As we have seen in Chapter 5, a person's Jewish status depends on his or her mother's religion (although Reform and Reconstructionist movements have accepted patrilineal descent, with certain conditions, in mixed marriages). Certain important prayers, including the *Mi-Shebeirakh* prayer for healing the sick, use a formulation requiring one's mother's name.

Among Orthodox and Conservative Jews, a male's tribal designation, Cohen, Levite, or Israelite, is passed down through his father. A Cohen's status has a legal bite. For example, according to Jewish law, a Cohen cannot marry a woman who is divorced or who was involved in a forbidden sexual relationship.[16] When called to the Torah, individuals are referred to as their father's offspring, as they are in Jewish ceremonies, such as weddings, and in Jewish legal proceedings.

Questions of permitted and forbidden marriages -- the range of relationships covered in the incest ban -- exist also. Under Jewish law, a man may marry his niece but not his sister. (In the United States, both unions are prohibited). If his sister were cloned, would the clone be considered like a niece or a sister? Thus, it is necessary to know one's parents to determine who is off limits to a cloned child under the incest prohibitions, briefly summarized in Chapter 6.

All this takes us back to Genealogy 101 under Jewish law. Who then are a clone's mother and father?

Let us start with maternity. Generally speaking, one's birth mother is one's mother under Jewish law.[17] As discussed in Chapter 6, a child conceived in vitro and implanted in a surrogate is the child of the woman who gives birth to him or her. Thus, if the nucleus of a mature cell is taken and implanted in that same woman's egg and after initial cellular divisions, the egg is returned to her uterus, the woman is the halakhic mother. She contributed the genetic material, carried the fetus to term, and gave birth. In this situation, no male directly contributed any DNA to this female's creation. Because the girl was delivered by a Jewish mother, the newborn would be Jewish.

According to some authorities, depending on the circumstances, a child may have two mothers, each of whom is considered his or her mother.[18] Under this interpretation, the woman who contributes the genetic material is one mother; the egg donor-surrogate, who contributes the mitochondrial DNA containing the encoded information that generate a variety of proteins (or protein portions) and carries and delivers the child, is another mother. There is a possible third mother if one woman contributes the genetic material, a second is the egg donor, and another serves as the gestational mother. A possible fourth mother also exists if embryo fusion brings together genetic material from each of two women, for instance, a lesbian couple.

Returning to the basic premise, if the gestational mother is presumed to be the halakhic mother, then, according to tradition, the child of a Jewish birth mother is Jewish. If the genetic donor is a non-Jewish woman, but the egg comes from a Jewish woman who bears the child, the child is Jewish. In the unlikely event that the Jewish tradition were to say that the genetic mother defines the child's status, then a child with a non-Jewish female genetic donor would not be Jewish.

Conversely, if the donor of the genetic material is Jewish and the gestational mother is non-Jewish, the child is not Jewish. The same conclusion follows if the egg donor-gestational mother is non-Jewish, even if the donor of the genetic material is Jewish (again, unless Jewish law looks to the genetic provider).

Because of the complexities brought about by modern technology, a child's Jewish status may be subject to doubt. In such a situation, traditionalists recommend that the clone-child undergo conversion to Judaism.

Now, let's consider paternity. If a husband contributes the genetic material and his wife provides the egg and serves as the birth mother, the wife is the halakhic mother. However, does halakhic fatherhood depend on the donation of the nucleic genetic material or being the sperm donor? As discussed in Chapter 6, a man who contributes sperm to donor assisted insemination is the father under Jewish law, at least for purposes of priestly status. In the context of cloning, the contribution of the nucleic genetic material by a male may be sufficient for him to be labeled the father for Jewish law purposes.[19]

What if no male genetic material were involved in the cloning process? Assume that the genetic provider is a Jewish woman. The nucleus is taken from her mature cell and is implanted in her egg. After the initial divisions occur, the egg is returned to the same woman. As we have seen, the woman who contributes the genetic material and gives birth to the fetus is the halakhic mother.

In this example, however, who is the clone-child's father? As one possibility, the clone could be the daughter of her mother's father (her paternal grandfather) because he supplied the male genetic materials to the mother's cells. By contributing to the somatic (non-reproductive cell) nucleus, but not making a sperm donation, his contribution is indirect. Thus, probably, in this situation, the female clone-child may have no father. Or, if the clone's birth mother is married, her husband could be the legal father of any child she may bear, through whatever means, as a result of marriage.

In dealing with these difficult parentage quandaries, recall the approach of Spiritual Judaism, developed in Chapter 5. Spiritual Judaism recommends forgetting about lineage and blood lines. If you say you are Jewish, that is sufficient to establish your religious status, subject to certain minimal conditions, set forth in Chapter 5.

Adultery: One Additional Jewish Concern

In addition to these identity and parentage issues, the Jewish tradition is also concerned about whether a child is born of an adulterous relationship. If an unrelated Jewish male (or a non-Jew) contributes the nucleic genetic material and the egg donor is a married Jewish woman, who carries and gives birth to the child, some traditionalists may view this reproductive process as involving adultery. They may reason that it is similar to the artificial insemination of a married Jewish woman by a Jewish (or a non-Jewish) man, not her husband. With cloning, Spiritual Judaism concludes, however, that the process should not be considered adulterous because sexual intercourse does not occur.

A More General Perspective on Cloning

Apart from these specific Jewish concerns, we need to consider three more general objections to human cloning, namely, the cloning of humans is repugnant and immoral, it is contrary to human dignity, and it will exacerbate social inequities.[20]

Human Cloning is Not Repugnant or Immoral

Critics of human cloning offer fuzzy misgivings, an unease, if you will, couched in terms of its being repugnant or repulsive.[21] They are uncomfortable, to put it mildly, with a technique for creating a child that is a younger, nearly identical genetic twin of another person. The uneasy reaction to cloning may be based, in part, on the initial images of clones, as slaves of their creator, in the popular media. The root of their disquiet most likely turns on the specialness of heterosexual reproduction by a married couple, through intercourse, joining a wife's egg with her husband's sperm in her body. The Jewish tradition also wants to safeguard the nuclear family unit, consisting of a husband-father and wife-mother, who establish offspring in a natural manner.

For Spiritual Judaism, cloning represents another in an array of assisted reproductive techniques, considered in Chapter 6. It is one among a number of procreative options, albeit an asexual method. Spiritual Judaism is willing to slide down the slippery slope of assisted practices deviating from our traditional concept of reproduction, including medical treatments for infertility, artificial insemination (combining a donated sperm with a woman's egg in her womb), in vitro fertilization (accomplishing the fertilization of an egg outside a woman's body), and surrogate motherhood (artificially inseminating a woman who nurtures a fetus to birth and turns it over to another individual or couple). Cloning represents a step beyond these technological advances that have changed our notion of procreation. Like these modern, assisted reproduction techniques, cloning involves a technological means to initiate the development of a fetus. It differs from nature only in technique, not in substance. Cloning will eventually come to be accepted as another form of assisted reproduction.

Because the Jewish tradition generally allows technological interventions in the reproduction process, individuals ought to have another option to the current alternatives. Once established as safe and effective, Spiritual Judaism recommends leaving the choice of a reproductive method to the involved adults (or adult), who may design their own arrangements and relationships. Human cloning does not pose significant harm to the clone or to others so as to contravene the important value of personal choice, exercised responsibly and in a loving manner.

Cloning is Not Contrary to Human Dignity

Opponents of human cloning view the practice as contrary to the rather vague concept of human dignity. For instance, Article 11 of the Universal Declaration on the Human Genome and Human Rights, promulgated in November, 1997 by the General Conference of the United Nations Educational, Scientific and Cultural Organization, as endorsed by the United Nations General Assembly in December, 1998, provides: "Practices which are contrary to human dignity, such as reproductive cloning of human beings, shall not be permitted."[22]

The human dignity argument appears to have two basic strands: first, cloning jeopardizes the personal identity and autonomy of the clone who will lack individuality and uniqueness; and second, cloning turns humans into commodities, artifacts, and objects thereby demonstrating a callousness toward and a depersonalization of human life.

In examining the first aspect of the human dignity approach, we need to ask what will be the impact of the process on a clone-child's psychological development? Will a clone encounter some sort of emotional deprivation or distress, a loss of a normal sense of self, or the freedom to create his or her own identity, if he or she lives in the shadow of a genetic progenitor? Will cloning deprive a person of his or her individuality, triggering what one Jewish commentator describes as the "ultimate 'identity crisis'."[23]

Spiritual Judaism rejects the genetic determinism implicit in the first strand of the human dignity argument. Prospective parents may opt to clone so as to provide a better genetic future for a child than offered by random genetic combinations. By cloning, they have the chance to give their offspring the genes of someone they admire, for instance, someone who lived into his or her nineties, free of genetic or neurological diseases or defects, with a zest for living. Yet, indeterminacy remains and the clone-child's future remains open.

The clone and his or her genetic ancestor will not be identical. Even in terms of their genetic composition, unless nuclear transfer cloning process involves the same female nucleus provider and egg provider, the clone and the cloner will have less than total genetic identity. Having nearly the same genes will not unduly restrict a clone's freedom, autonomy, or individuality. The clone and the cloner will have to come to term in different uteruses. These differences in uterine environments will contribute to divergent personalities and temperaments. The genetic predecessor will have been born a generation earlier and have grown up in a different world. Clone and cloner will grow up in disparate environments, encounter divergent life experiences, personal relationships, peers, and enjoy differences in diet and healthcare. Their emotional development and human characteristics, such as their values, will vary. Random events, luck and chance, as well as life choices and freewill, all will play a role in the development of the clone's personality, character, behavior and intellectual capabilities. Nature -- one's genes -- and nurture -- one's physical, social and psychological, and moral environment and experiences -- will shape a clone in a multitude of ways and differentiate him or her from his or her genetic ancestor. Environmental factors and external stimuli may depress (or activate) certain genetic tendencies. We may hope that a clone will learn from his or her genetic progenitor's mistakes. Last, but not least, the clone and his or her genetic ancestor will have different souls.

Two types of empirical evidence bear out the conclusion that clones will not face emotional difficulties. First, test-tube babies, the products of in vitro fertilization, do not seem to have encountered high rates of psychological problems.[24] A clone should not have more questions about his or her self-worth and identity, among other psychological issues, than a test tube baby.

Second, genetically identical twins, who have the same nucleic and non-nucleic DNA, are not deprived of their individuality. Born with different neural tangles, they develop different psychological and personal characteristics, temperaments, mental processes, and life choices, stemming from their environmental experiences, including life histories, personal relationships, and their own free choices.[25] Identical twins, despite their social closeness, possess a considerable degree of individuality; they develop separate personalities. Divergent life experiences cause unique brain circuitry to form in each. Even identical twins with exactly the same genetic makeup and a somewhat similar environment have separate souls.

Clones will not lose the possibility of freedom and spontaneity merely because of their mode of reproduction. From what is known about identical twins, cloning will not reduce human free will. Clones will still be able to actualize their own potentialities, even if their genetic ancestors possess somewhat similar capabilities. Clones will not feel that their future is substantially preordained or that their emotional growth is stunted. The character development of cloned and non-cloned children takes many years. It is subject to numerous unexpected environmental influences. Humans are quite adaptable; a cloned child likely will not worry about his or her identity.

The second aspect of the human dignity argument focuses on looking at a clone as an object of manipulation by others, thereby leading to a lessening of the worth of persons in general and a diminished respect for human life. The fear is that a clone, particularly one created using genetic material from someone meaningful to the user of this process or someone of extraordinary talents and accomplishments, would be valued as a commodity to be created and enjoyed, not as a person or because of his or her "intrinsic" worth or value.

Opponents of human cloning offer a number of scenarios as raising a threat to the ideal parent-child relationship and what it means to raise a child. Most basically, cloned children might be treated as a means to various ends sought by parents, not as ends in themselves. For example, parents could clone a terminally ill child who accidentally suffered a brain or other fatal injury to produce a younger, nearly genetically identical twin. In this case, apart from the question of the child-cloner's inability to consent to the process, the clone might be a living memorial to someone else; however, so might non-cloned children. Parents may also want a clone whose tissue cells will be available for transplant to save an older sibling who, it is asserted, may be the real focus of parental love. Or, the narcissistic, self-absorbed could strive to reproduce themselves in their own image thereby attempting to attain a measure of immortality.

The difficulty with the objectification argument turns on the reality of parent-child relationships. Today, as in the past, children are conceived and brought up for a number of less than ideal reasons, including being accidents, not intended at all and perhaps not wanted, serving as instruments of parents' thwarted ambitions, enabling parents to live vicariously through them, or reflecting a parental desire for immortality or the search for caregiving and companionship in old age. In short, children, whether cloned or not, may serve as instruments of parental fulfillment.

Even with sexual reproduction, parents may exercise excessive control over their children not only thwarting their independence but also turning them into objects. Children often face excessive parental pressures to attain high accomplishments. A clone-child will not face any more unreasonable expectations than those of other parents who hope (if not expect) a child to achieve in his or her lifetime what they have been unable to attain. At the end of the day, nearly all parents will love a child, cloned or not, no matter what he or she does or achieves.

Human Cloning Will Not Unnecessarily Exacerbate Social Inequalities

Human cloning and its closely related technological process, genetic enhancement (as we will see in the next chapter), may exacerbate social inequalities. In reproducing, whether sexually or asexually, the wealthy may have easier access, at least so the suspicion runs, to these techniques, while the less affluent must continue to depend on chance. A concern also exists about the possible development of a black market for cloned embryos from people with "desirable" traits, such as intelligence or good looks, who would be sold to potential parents.

All appropriate steps must be taken to prevent such abuses. These having been provided, Spiritual Judaism advises us not to allow our fears of the distributional consequences to limit technological advances, including human cloning. First, the specter of social inequities is congruent with other choices modern capitalistic society allows parents. Today, affluent parents can offer their children all sorts of environmental enhancements, from summer camps to latest computer technology, from music lessons to a first rate college education. Second, the cost of these technological advances will likely drop substantially in the future and come within the reach of average income or even poor couples (or individuals). Third, some of these modern techniques may become governmental entitlements, available to all, or may be covered by insurance.

Primary guidance and regulation of all research and clinical practice with respect to cloning should be in the hands of scientific, medical and counseling professionals. Spiritual Judaism suggest that rabbis keep abreast of developments and inform interested congregants, together with discussions of the implications and the ethical issues. As for governments, at any level, while prudence requires that they also keep a close watch and act to prevent harm to people, let them support the development of knowledge and safe clinical practices. In other words, the public sector, the federal government, the various states, and transnational bodies, ought to adopt a hands off policy with respect to human cloning. Spiritual Judaism recommends that governments not impose any bans or regulations on either research involving human cloning for valid scientific or medical purposes (or its funding, by public or private sources) or, ultimately, on the cloning of humans. Through accumulated studies of nonhuman mammalian cloning, the safety and effectiveness issues regarding the cloning of human beings will become clearer. In the meantime, let us leave the regulation of human cloning to hospital ethics committees and local review committees, known as institutional review boards. These decentralized, nongovernmental bodies are more readily able to take into account new developments and studies of nonhuman mammalian cloning than legislators or governmental bureaucrats.

The permissive conclusion offered by Spiritual Judaism will, however, necessitate some minor statutory modifications. Governments will need to extend the existing ban on commercial transactions involving embryos, fetuses, and babies (as well as human organs) produced by sexual reproduction to those generated by cloning. Although the question of who is related to whom could be handled by contract, a need also exists for the various states in the American legal system to devise a statutory definition for the parentage and inheritance rights of clones. In addition to delineating who are a clone's parents, it is important to answer the question of whether a clone has a right of inheritance from the person who donated the nucleic genetic material. Also, legislators ought to clarify whether parents (or a single parent) may provide the consent to clone a minor child.

Because of the availability of alternative means of reproduction discussed in Chapter 6, cloning will likely figure only marginally in future as a procreation technique. However, for those infertile couples, who cannot currently benefit from existing assisted reproductive techniques, it represents the only way to have biological children. Furthermore, perfecting cloning will help make genetic engineering, the process of adding or altering genes, a reality, a topic we turn to in the next chapter.

Notes

1. Sheryl Gay Stolberg, "Despite Warnings, 3 Vow To Go Ahead On Human Cloning," *New York Times* August 8, 2001, A1; Rick Weiss, "Scientists Declare Progress on Human Cloning," *Washington Post*, August 8, 2001, A2; Christian Science Monitor, "Quests Build to Clone Humans," February 7, 2001, 1; Margaret Talbot, "A Desire to Duplicate," *New York Times Magazine*, February 4, 2001, 40; Rick Weiss, "Human Cloning's 'Numbers Game,'" *Washington Post*, October 10, 2000, A1; Rick Weiss, "Embryo Work Raises Specter of Human Harvesting: Medical Research Teams Draw Closer to Cloning," *Washington Post*, June 14, 1999, A1; Michael Waldholz, "Cloning of Humans Grows Increasingly Possible," *Wall Street Journal*, December 18, 1998, B12; Sheryl WuDunn, "South Korean Scientists Say They Cloned a Human Cell," *New York Times*, December 17, 1998, A12.
2. According to some experts, few cloned humans will survive to birth and of these, the majority may be abnormal. Rudolf Jaenisch and Ian Wilmut, "Don't Clone Humans!," *Science* 291:5513 (March 30, 2001): 2552. See also Gina Kolata, "Researchers Find Big Risk of Defect in Cloning Animals," *New York Times*, March 25, 2001, A1.
3. I. Wilmut et al, "Viable Offspring Derived from Fetal and Adult Mammalian Cells," *Nature* 385:6619 (February 27, 1997): 810-813. Gina Kolata in *Clone: The Road to Dolly and the Path Ahead* (New York: William Morrow, 1998) describes the science leading up to Dolly. See also Michael Specter and Gina Kolata, "After Decades and Many Missteps, Cloning Success," *New York Times*, March 3, 1997, A1.
4. Todd S. Purdum, "President Asks Experts for Advice on the New Reality of Cloning," *New York Times*, February 25, 1997, A15.
5. Remarks by the President Announcing the Prohibition on Federal Funding for Cloning of Human Beings, March 4, 1997, Weekly Compilation of Presidential Documents 33:10 (March 10, 1997): 278-279; Memorandum on the Prohibition on Federal Funding for Cloning of Human Beings, *Id.*, at 281; Marlene Cimons and Jonathan Peterson, "Clinton Bans U.S. Funds for Human Cloning Research," *New York Times*, March 5, 1997, A1.

6. National Bioethics Advisory Commission, *Cloning Human Beings: Report and Recommendations of National Bioethics Advisory Commission* (Rockville, MD: National Bioethics Advisory Commission, 1997), 108-109.
7. Message to Congress Transmitting the Proposed "Cloning Prohibition Act of 1997," June 9, 1997, *Weekly Compilation of Presidential Documents* 33:24 (June 13, 1997): 845-846; Draft Legislation Entitled the "Cloning Prohibition Act of 1997," 105th Congress, 1st Session, House Document 105-97.
8. The Bush Administration has indicated that it opposes all forms of human cloning, including therapeutic cloning for medical research. Antonio Regaldo, "Bush Administration Says It Opposes Cloning of Human Embryos for Research," *Wall Street Journal*, June 21, 2001, A4; Rick Weiss, "Bush Backs Broad Ban on Human Cloning," *Washington Post*, June 21, 2001, A1. In July, 2001 the U.S. House of Representatives voted for a sweeping ban on human cloning both to create individuals and to conduct research or treat disease. Sheryl Gay Stolberg, "House Backs Ban on Human Cloning For Any Objective," *New York Times*, August 1, 2001, A1; Rick Weiss and Juliet Eilperin, "House Votes Broad Ban On Cloning," *Washington Post*, August 1, 2001, A1.
9. California Codes, Health and Safety Code §24185; Louisiana Revised Statutes, Title 40, §1299.36.2; Michigan Compiled Laws, §750.430a; General Laws of Rhode Island §23-16.4-2.
10. Federal Register 58:197 (October 14, 1993): 53248-53251.
11. Federal Register 66:13 (January 19, 2001): 5447-5469.
12. The Ethics Committee of the American Society for Reproductive Medicine, "Human Somatic Cell Nuclear Transfer (Cloning)," *Fertility and Sterility* 74:5 (November 2000): 873-876, at 875.
13. I have drawn on John D. Loike and Avram Steinberg, Human Cloning and Halakhic Perspectives, Tradition 32:3 (Spring 1998): 31-46; Avraham Steinberg, "Human Cloning-Scientific, Moral and Jewish Perspectives," The Institute for Jewish Medical Ethics of the Hebrew Academy of San Francisco, www.ijme.org; Michael Broyde, "Cloning People: A Jewish Law Analysis of the Issues," *Connecticut Law Review* 30:2 (Winter 1998): 503-535; Michael J. Broyde, "Cloning People and Jewish Law: A Preliminary Analysis," *The Journal of Halacha and Contemporary Society* 34 (Fall 1997):
27-65; Jonathan R. Cohen, "In God's Garden: Creation and Cloning in Jewish Thought," *Hastings Center Report* 29:4 (July/August 1999): 7-12, reprinted in *The Human Cloning Debate*, 2nd edition, ed. Glenn McGee (Berkeley, CA: Berkeley Hills Books, 2000), 251-267; Elliot N. Dorff, *Matters of Life and Death: A Jewish Approach to Modern Medical Ethics* (Philadelphia: Jewish Publication Society, 1998), 310-324; Elliot N. Dorff, "Human Cloning: A Jewish Perspective," *Southern California Interdisciplinary Law Journal* 8:1 (Winter 1998): 117-129.
14. William N. Eskridge, Jr. and Edward Stein, "Queer Clones," in *Clones and Clones: Facts and Fantasies About Human Cloning*, ed. Martha C. Nussbaum

and Cass R. Sunstein (New York: Norton, 1998), 99, briefly discuss the possibility of this feminist utopian notion.
15. Charlotte J. Patterson, "Adoption of Minor Children by Lesbian and Gay Adults: A Social Science Perspective," *Duke Journal of Gender Law & Policy* 2:1 (Spring 1995): 191-205, at 191.
16. Rabbi Shlomo Ganzfried, *Kitzur Sholchon Oruch*, volume II, trans. Rabbi Eliyahu Touger (New York: Moznaim, 1991), Chapter 145:12, The Laws of Marriage, 607.
17. Broyde, "Cloning People: A Jewish Law Analysis", 511-513, 515, 519, 533-534; Broyde, "Cloning People and Jewish Law", 33-36, 38, 63. See generally, Michael Broyde, "The Establishment of Maternity and Paternity in Jewish and American Law", *National Jewish Law Review* 3 (1988): 117-158.
18. J. David Bleich, "Cloning: Homologous Reproduction and Jewish Law," *Tradition* 32:3 (Winter 1998): 47-86, fn. 71, at 83-84; J. David Bleich, "In Vitro Fertilization: Questions of Maternal Identity and Conversion," *Tradition* 25:4 (Summer 1991): 82-102, at 86-88. See also Broyde, "Cloning People: A Jewish Law Analysis," 512-513, 515; Broyde, "Cloning People and Jewish Law," 34-36, 38-39.
19. Broyde, "Cloning People: A Jewish Law Analysis," 513, 515-516, 519; Broyde, "Cloning People and Jewish Law," 43-44.
20. I have drawn on *Cloning Human Beings*, 62-84 and Dan W. Brock, "Cloning Human Beings: An Assessment of the Ethical Issues Pro and Con," in *Clones and Clones*, 141-164; Carson Strong, "Cloning and Infertility," *Cambridge Quarterly of Healthcare Ethics* 7:3 (Summer 1998): 278-293.
21. Leon R. Kass, "The Wisdom of Repugnance," The New Republic, June 2, 1997, 17-26, at 19-21; Leon R. Kass, "The Wisdom of Repugnance," in *The Ethics of Human Cloning*, eds. Leon R. Kass and James Q. Wilson (Washington, D.C.: The AEI Press, 1998), 17-19. See also Leon R. Kass, "Why We Should Ban Human Cloning Now: Preventing a Brave New World," *The New Republic* (May 21, 2001): 30-39.
22. United Nations Educational, Scientific and Cultural Organization, Twenty-Ninth Session of the General Conference, Universal Declaration on the Human Genome and Human Rights, Article 11, adopted on November 11, 1997; Resolution 53/152, United Nations Resolution on the Human Genome and Human Rights, adopted by the United Nations General Assembly on December 9, 1998, A/RES/53/152(10 March 1999).
23. David M. Feldman, *Birth Control in Jewish Law: Marital Relations, Contraception, and Abortion As Set Forth in the Classic Texts of Jewish Law* (Northvale, N.J.: Jason Aronson, 1998), 331.
24. James Q. Wilson, "Sex and Family," in *The Ethics of Human Cloning*, 94-95.
25. Nancy L. Segal, "Behavioral Aspects of Intergenerational Human Cloning: What Twins Tell Us," *Jurimetrics* 38:1 (Fall 1997): 57-67, at 63-65.

Chapter 8

The Road to Designer Children: Sex Pre-selection, Genetic Testing, and Genetic Engineering

Reproduction, as fundamental to humans as it is, has become, as developed in Chapter 6, increasingly technology-driven. From conception through birth it is hard to find an area where technology has not had an impact. Beginning in the 1960s physicians began to use ultrasound (sound waves to create images of organs) to examine fetuses in the womb, thereby observing prenatal development and diagnosing abnormalities. With the 1970s came more technological gymnastics, notably IVF letting infertile couples conceive. The following decades have witnessed numerous other tests and treatments for the unborn, most recently, genetic therapies and fetal surgeries. Someday, physicians will diagnose genetic predispositions to disease, treat ailments before they occur, and correct those occurring at the genetic level.

These technological advances often leave couples facing difficult decisions for which they are emotionally unprepared. Increasingly, couples expect perfect babies all the time, leading to a desire for designer-made-to-specification -- "perfect" -- children.

From the perspective of the Jewish tradition, technological advances prompt us to ask: when do humans have a right to take the place of God? Modern technologies, including sex pre-selection, genetic screening and engineering represent a new power to control human destiny. As the late Lord Jakobovits warned: "...'[G]enetic engineering' may open a wonderful chapter in the history of healing. But without prior agreement on restraints and the strictest limitations, such mechanization of human life may also herald irretrievable disaster resulting from man's encroachment upon nature's preserves...and from replacing the matchless dignity of the human personality by test-tubes, syringes and the soulless artificiality of computerized numbers."[1]

This chapter examines what the Jewish tradition has to say about imposing limits on the gratification of parents' desires, when in the future technology will allow parents to fulfill their wish for a "perfect" baby. Should we fashion children to specification? Should we use every new technology? Should we preserve the elements of life's mystery? Should we attempt complete, or, at least as effective as possible, control over the outcome? Should we try to defeat aging and death? How can we enrich the human condition yet continue to exalt the Eternal?

In answering the questions of what is God's Domain and is it wrong to tread on the Eternal's Territory?, the mainstream of Judaism often sounds a cautionary note, reminding us of the warning coming from the tale of the Tower of Babel. The author of the Book of Genesis provided a commentary on human hubris, the overstepping of appropriate bounds, in the attempted construction of the Tower of Babel (Genesis 11:1-9). At Babel, humans wanted to show that they had the capacity to reach to the level of God. They set out to build a tower that would rise up to heaven itself. The Eternal, we are told, preferring that humans stay within the earthly domain, halted the structure by confounding language and dispersing people across the globe. For some traditionalists, the moral of the story is clear -- do not change the "natural" order of things.

This chapter offers a much more positive note. God gives to us the tools and the knowledge to tamper with the Divine arrangement of the world. The Jewish tradition teaches that we ought to cure, and if possible prevent, illnesses and diseases, regarded as harmful evils to be overcome as effectively as possible. We can aid God in the Divine activity of healing and even preventing ill health. Yet, we stand on the verge of controlling our own destiny, a position never before occupied by humans. The mainstream of the Jewish tradition welcomes genetic testing, therapeutic genetic engineering, but not sex pre-selection or genetic enhancement. Spiritual Judaism goes further than this by approving sex pre-selection and genetic enhancement. Spiritual Judaism suggests that it is safe and right to leave to individuals, not to government regulation or religious proscriptions, the power to decide and implement what they regard as a "good" change, what they see as "normal" or desirable.

Sex Pre-selection Techniques

Sex pre-selection involves interventions prior to and after conception designed to predetermine the sex of a child. Today, physicians use pre- and post-conception techniques to obtain offspring of a desired sex. As the availability of and interest in reliable sex pre-selection techniques grows, more couples may seek to ensure a child of a desired sex.

Physicians employ one of several strategies. Sperm separation, a pre-conception technique, involves the separation in a laboratory of sperm into androsperm (male-bearing) and gynosperm (female-bearing).[2] The desired sperm type is then artificially inseminated into a woman. One post-conception technique combines IVF with genetic testing to ascertain the sex of the four-cell embryo prior to implantation in the womb. At present, IVF is far more expensive than sperm separation, making sperm separation the sex pre-selection method of choice.

Sex Pre-selection in the Jewish Tradition

Jewish sages ardently pursued the topic of sex selection. The Talmud records no opposition by the rabbis to the notion of various types of interventions to affect the sex of offspring. Letting wives take their pleasure first will produce boys as we read in the Babylonian Talmud (Niddah 31a-b):

> Rabbi Isaac, citing Rabbi Ammi, stated: If a woman [has an orgasm] first she bears a male child; if a man [has an orgasm] first she bears a female child; for it is said, "If a woman emits and bears a man-child" (Leviticus 12:2).
>
> Our rabbis taught: At first it used to be said that "if the woman [has an orgasm] first she will bear a male, and if the man [has an orgasm] first she will bear a female," but the sages did not explain the reason, until Rabbi Zadok came and explained it: "These are the sons of Leah, whom she bore to Jacob in Paddan-aram, with his daughter Dinah (Genesis 46:15), the Hebrew Bible thus ascribes males to the females (the sons of Leah) and the females to the males (Jacob's daughter Dinah).
>
> "And the sons of Ulam were mighty men of valour, archers; and had many sons, and sons' sons" (I Chronicles 8:40). How is it within the power of man to increase the number of "sons and sons' sons"? But the fact is because they contained themselves during intercourse, in order that their wives should

> [have orgasms] first, so that their children shall be males, the Hebrew Bible attributes to them the same merit as if they had themselves caused the increase in the number of their sons and sons' sons. This explains what Rabbi Kattina said, "I could make all my children to be males." Raba stated: "One who desires all his children to be males should cohabit twice in succession."

The Babylonian Talmud contains a number of other suggested techniques to attain the desired births of males.[3] One sage recommended that the orientation of the marital bed in a north-south direction would produce a male child (Berakoth 5b). Some suggested giving charity to the poor or making one's wife happy prior to intercourse (Baba Bathra 10b). Still others focused on the observance of the havdalah ceremony at the end of the Sabbath or the sanctity with which intercourse is performed (Shevuoth 18b). Finally, to have male children, some advised marrying a wife worthy of one's self and conducting one's self with modesty at the time of intercourse (Niddah 70b-71a).

For contemporary traditionalists, the question of sex selection focuses not only on the textual exegesis but also on the technique used. As Modern Orthodox Rabbi J. David Bleich states, "The primary halakhic concern is not with regard to the decision to engage in sex preselection but the method to be employed in effecting sex determination."[4] Although discouraging sex pre-selection, preferring natural procreation, Bleich does not oppose the process, provided halakhically approved methods are used. Because sex determination is not a legitimate grounds for pregnancy termination, Bleich, however, rules out abortion if a fetus is not of the desired sex.[5]

Traditionalists prefer IVF to sperm separation, but stringent limitations exist on the use of IVF. Because IVF involves a dramatic interruption of natural marital intercourse, as we have seen in Chapter 6, traditionalists require lengthy post-marriage waiting periods of at least two years before allowing the use of IVF. Thus, IVF can be used only by infertile couples unable to conceive by other means.

It is unclear whether traditionalists would permit the use of IVF for the purpose of sex selection in the absence of marital difficulty in conceiving.[6] Some would view the securing of a desired gender child as too frivolous to be halahkically acceptable. For them, the imperative to maintain natural marital intercourse remains paramount.

Because untransplanted embryos lack standing in Jewish law, if IVF is properly used to overcome infertility problems, it is then permissible to transplant only fertilized eggs of the desired sex. If untransplanted eggs fertilized in vitro may be discarded, then sperm in excess of that needed to fertilize eggs can also be discarded, seemingly without violating the prohibition of the wasting of seed so sacred to traditionalists.

Not unexpectedly, objections exist to sperm separation. Most traditionalists reject assisted insemination by husband (AIH), unless medically indicated after a waiting period following marriage of two to eight years so as to establish medical necessity. Thus, only couples with a long history of fertility difficulties can use AIH. Furthermore, AIH is permitted only if none of the husband's sperm is discarded prior to fertilization. Because sperm separation involves the deposit of either androsperm (male-producing) or gynosperm (female-producing) in a male's wife, the discarding of the unused sperm would constitute, for traditionalists, the prohibited destruction of seed.[7]

A Liberal Approach to Sex Pre-selection

The Reform Movement exhibits considerable ambivalence with regard to sex pre-selection. A 1941 Reform responsum dealing with this topic stated that according to Jewish sages, "The desire of parents to predetermine, if possible, the sex of their progeny, is not a reprehensible desire. The objective sought is a legitimate objective." Taking a permissive position, the responsum concluded, "[I]f the means... will prove scientifically sound and morally unassailable, the Jewish teachers... will find ample basis for their endorsement of the enterprise in the thought and tradition of their past."[8]

One contemporary Reform commentator, Rabbi Daniel Schiff, sounds a far more cautionary note. Harkening back to what he (and traditionalists) regard as "natural," Schiff reasons that Judaism sees something holy about conceiving children within the context of marital sexual intercourse, the most sanctified means of procreation. Schiff states, "Quite simply, [the production of children through marital sexual intercourse] is a hallowed idea." Moving from this premise, leads Schiff, in contrast to the positions of liberal Reform responsa set forth in Chapter 6, to view AIH and IVF as unnecessary diminutions of the "sanctity and [holiness] accompanying the marital union." Because sperm separation and IVF break the nexus between marital sexual intercourse and conception, Schiff frowns on these techniques to achieve gender determination. In particular, he rejects the IVF technique, even under the conditions of the 1941 responsum, on the grounds of complexity and safety concerns.[9]

The Search for a Sounder Approach to Sex Pre-selection

Beyond interpreting ancient texts, modern commentators, such as Rabbi Schiff, raise much deeper concerns. They conclude that sex pre-selection is invidious and evidences a gender bias, thereby reinforcing discriminatory attitudes toward women and diminishing the status of females.[10] For these critics, sex pre-selection is seen as implicitly fostering the value of one sex over another and confirming that sex represents a governing factor in human behavior. They rely on studies showing a marked preference for male children in general and particularly for male first-born. Looking to its social impact, they reason that efficient sex pre-selection may lead to social upheaval, specifically, a gender imbalance resulting from an oversupply of males or at least a higher preponderance of first-born males, yielding in turn, a lack of marriage partners and a possible increase in rape and other acts of violence against women.

Fears that sex pre-selection will result in a deficiency of females may be overblown. We should not condemn this technique on the basis of speculation[11] and the absence of proof in Western societies. Some experts view the likelihood of an unbalanced sex ratio as a "very remote possibility."[12] If couples choose boys in excessive numbers, they may face social disapproval.

Spiritual Judaism urges respecting the reproductive freedom of couples (and singles) who may want to use sex pre-selection to balance their family in gender, thereby avoiding the financial strain of children before the boy or girl they want arrives. They may also seek to provide a better quality of life for a child for the wanted sex. Finally, even after sex pre-selection women can always refuse to stay pregnant, a topic to be considered in detail in Chapter 9. In short, Spiritual Judaism suggests letting parents go ahead, if they want and exercise control over a child's gender.

But why stop at gender selection, what about disease prevention to ensure better health and then body size and height? Before discussing these questions, let us first consider the Human Genome Project and genetic testing.

Genes and Proteins: Current Realities and Future Possibilities

Advances in molecular biology and biochemistry have revolutionized our understanding of the work within the cell, the fundamental unit of life, that stores, transmits and edits information and then uses that information to regulate fundamental biological processes.[13]

In the Human Genome Project,[14] scientists mapped the 30,000 (or more) genes[15] that store genetic information in the human body. Genes are basically sets of instructions directing the body to make proteins. Proteins, the molecular building blocks of life and the workhorses of the body, perform numerous functions ranging from fighting disease to giving human tissues their structure. Although scientists are learning to read the genetic programming instructions shaping, in part, human beings, much work remains to be done in cracking the mystery of proteins and discovering the body's collection of proteins, known collectively as the "proteone." The tasks ahead include how a gene serves as a script directing the making of a protein (or proteins), what a protein does, and how it interacts with other proteins. Genomics is important, but proteins do the work. The links between a gene, a protein (or proteins), and a disease are complex and indirect. Figuring out how the inner machinery of cells work, how malfunctions lead to disease, and how to devise new diagnostic tests and make new drugs, rests on a more thorough understanding of proteins.

For some diseases at least a genetic connection can now be isolated with a reasonable degree of certainty. This knowledge will help researchers and physicians who have long known the existence of deleterious genes, manifesting themselves in serious diseases and defects.

There are inherited, clearly defined, genetic diseases. One example is Tay-Sachs, a rare, horrific genetic disease, clustering among Ashkenazi Jews of Central and Eastern European descent. Tay-Sachs results from a baby's inability to produce one critical enzyme needed for normal brain growth and development. The disease is totally fatal. Newborns appear normal; however, after several months they start to exhibit severe muscular weakness. They evidence difficulties in doing activities infants typically do, rolling over, crawling, walking, swallowing, and ultimately, breathing. Blindness, seizures, and a haze of unresponsiveness compound these problems. Progressive motor and mental deterioration occur before death inexorably results at between the ages of three to five.

Through our genes (and the proteins they make), we also carry susceptibility to various other diseases, potentially causing debilitating or deforming illnesses or conditions. Our genes put us at greater risk to certain physical diseases, such as cancer, cardiovascular problems, and arthritis, as well as mental disorders, such as depression and schizophrenia. Various kinds of behavior, such as alcohol and drug abuse, may be genetically pre-disposed.

The mapping of the human genome and in the future, an understanding of proteins and the basic biochemical mechanisms underlying sickness and health, open the door for the early diagnosis of thousands of disorders, ranging from mental conditions to heart disease. Today, genetic information about one's health and health prospects can be obtained by genetic testing. The isolation and characterization of human genes offers opportunities for disease detection, diagnosis, screening, prevention, treatment, and care by gene replacement or other types of therapy. However, we need to remember that most illnesses are influenced by a complex interaction of numerous genes and the proteins they produce, with most of these interactions currently far from being understood.

Genetic Testing: An Overview

Through pre-implantation embryo screening, physicians test for a mutant gene or gene product in IVF eggs. After a woman's eggs are removed from her ovaries and fertilized in vitro, a geneticist punctures each egg's coating. Using a glass needle, a single cell, one seventh the width of a hair, is sucked out of an eight-cell embryo.

The focus currently centers on genetically testing of one cell, known as, single-cell genetics, isolating one specific gene, and looking for the presence or absence of a genetic mutation for, among other conditions, Tay-Sachs disease, cystic fibrosis, hemophilia, muscular dystrophy, or Huntington's disease (a degenerative nerve disorder that usually does not appear until between the ages 35 to 50).[16] Screening is used by parents having a child suffering from a severe genetic illness or if the couple has tested positive for a specific genetic mutation.

Physicians discard the affected in vitro zygote(s) that do not pass genetic muster and implant into the mother's womb one or more embryos free of genetic disease. This technique helps facilitate the birth of a healthier baby, provided the implantation and the pregnancy are successful, thereby avoiding the physical and emotional trauma of abortion. In the IVF clinic, medical expertise controls the decision over which embryos to implant and which to discard.

Technological changes will make prenatal testing far easier. In the near future, technology will make it possible to collect genetic material for various tests from simple blood samples, isolating fetal cells floating in its mother's blood rather than the current invasive techniques, that involve inserting a needle to collect amniotic fluid (amniocentesis) or sample placental tissue (chorionic villus sampling). These new diagnostic tests will allow physicians to detect disease early when it can be treated, rather than when it is too late.

In coming decades, new tests, using DNA chips, will be capable of reading hundreds, if not thousands, of genes simultaneously, testing for many traits rather than one at a time, from one embryo cell. When an embryo's DNA bonds to a specific pattern, signaling an aberrant gene sequence, a scanner will read the matched pattern and feed the results into a computer that will spit out the embryo's genetic report card.

These sophisticated tests will present new and perplexing problems for parents. Genetic tests only indicate the risk or the odds for developing a condition or illness that may not arrive for years, not whether an individual will actually come down with that disease. The association between one's genetic makeup and the prognosis for disease remains statistical. For example, if a fetus has a breast cancer gene, ought you to abort because there is a chance it will develop the disease later in life? Parents must remember a genetic predisposition is just that, a predisposition, not a certainty. Some diseases are simply too random to predict with any certainty.

With respect to pre-implantation screening, most embryos will have some genetic defects. Which embryo to implant, the one with a likelihood for colon cancer or the one with a propensity for diabetes or breast cancer that may not click in for forty years, if at all? Or, should parents toss out all the embryos and start over?

Our ability to screen for diseases today outpaces medicine's ability to treat these conditions. We often lack the tools to replace or suppress a defective gene or drugs to mitigate its destructive impact.

Restoring Health and Preventing Genetic Problems in the Jewish Tradition

The Jewish tradition views as Divinely sanctioned, if not mandated, the acquisition of knowledge to cure human illnesses and disease. The pursuit of knowledge represents a legitimate activity for humans, not an encroachment on God's plan for this world. Specifically, the pursuit of medical knowledge does not involve the prohibited eating from the tree of knowledge (Genesis 2:17).

The Babylonian Talmud (Baba Kamma 85a) interprets the phrase "and heal he shall heal" (Exodus 21:19) as allowing the use of knowledge to cure disease and heal illness. This use of knowledge may even be biblically mandated based on the obligation to restore a lost object (Deuteronomy 22:2), including the restoration of lost health.

Rabbinic sources support the notion of avoiding hereditary diseases and genetic problems. One Jewish sage, Raba, advised a man against marrying a woman from a family of epileptics or lepers, provided the disease occurred three times in that family (Babylonian Talmud Yevamot 64b) so that the illness would not be genetically transmitted to future generations. Lord Jakobovits noted that this rabbinic ruling "may well represent the first eugenic enactment, and the only legislative bar to the procreation of a diseased progeny, in ancient and even medieval times."[17]

In another section of the Babylonian Talmud (Bekoroth 45b) we read that Resh Laskish, who was concerned that children who stood out from their peers would became objects of ridicule, advised: a tall man not to marry a tall woman lest they give birth to a giant; a male dwarf not to marry a female dwarf lest they give birth to an even smaller dwarf; a very pale man not to marry a very pale woman lest they give birth to an albino.

Implications of the Jewish Tradition for Genetic Testing

Two basic uses for genetic testing exist prior to birth. The first is therapeutic, that is, the curing or lessening the susceptibility to conditions or diseases. The second involves non-therapeutic gene manipulation.

There seems to be little or no objection, from the mainstream of the Jewish tradition, to genetic screening. Assuming they permit IVF, which involves an interruption of "natural" marital relations, traditionalists would approve the use of pre-implantation screening of in vitro fertilized eggs for Tay-Sachs disease, Huntington's disease, and hemophilia, among other maladies, choosing only unaffected zygotes for implantation, thus preventing the birth of a "defective" child. Discarding affected zygotes avoids the issue, for traditionalists, of pregnancy termination. Pregnancy and the status of the fetus as a potential human being in Jewish law begins only with zygote implantation into the wall of the uterus.[18]

Several practical downsides exist. Genetic testing adds to the cost of IVF, already an expensive procedure. Also, questions exist as to whether or not the screening is foolproof. The tests probing for genetic alteration provide information, predicting, with varying degrees of reliability, the odds of succumbing to a deadly disorder.

For traditionalists who oppose IVF, the route to genetic testing focuses on another technique, uterine lavage (UL). UL involves flushing out an embryo soon after fertilization through natural marital relations. The downside is the risk of an incomplete flush possibly resulting in an unscreened pregnancy. Because the flushing takes place before the implantation of an embryo and the commencement of pregnancy, traditionalists will likely permit this technique, especially if the embryo is retained for implantation.[19]

Post-implantation prenatal screening involves testing cells, amniotic fluid or placental tissue, to detect fetal abnormalities. Diseases transmitted as single-gene defects, such as Tay-Sachs disease and cystic fibrosis, can be diagnosed early in pregnancy. Tests also exist for Down Syndrome, a chromosomal disorder, and a propensity for breast cancer.

For traditionalists who adhere to a stringent position on abortion, as analyzed in Chapter 9, basically prohibiting abortion beyond forty days after conception, the problem arises if genetic testing is used for the purpose of influencing a decision to abort (and then aborting) a fetus carrying a genetic defect for a disease or condition for which no effective treatment currently exists. Beyond forty days post-conception, traditionalists condemn the abortion of a fetus, apart from maternal considerations, even one with a susceptibility to Tay-Sachs disease.[20]

Genetic Screening: Some Contemporary Considerations

Apart from the position of traditionalists on IVF and abortion, critics of genetic screening point to the developing culture of "choosiness" in which parents (and their physicians) pick embryos with the "best" genetic profile, eliminating embryos which may result in children with even minor, curable, or late onset conditions and diseases.[21] With parents becoming pro-active consumers, these opponents see a new, "unnatural" relationship evolving between parents and children. When a child's physical or mental makeup is overseen by his or her parents, he or she may become a selected product, expressing, in part, his or her parents' aspirations, desires, and values.

In responding to these objections, proponents of genetic screening note that parents want what they regard as "best" for their children, at least until their offspring reach a certain age and establish their independence. It does not seem likely that parents will love their children any more or less for who they are, regardless of their abilities or talents, whether the offspring result from the biological lottery or are culled to make them what the parents want.

Selecting an embryo or a fetus on the basis of the absence of genetic disease, it is asserted, could also increase negative attitudes toward the physically or mentally disabled. Active elimination of genetic diseases could also send a devaluing message to individuals having the condition or illness being eliminated.

These fears must not lead to a condemnation of genetic screening. Spiritual Judaism recommends allowing individuals, on a couple-by-couple or single-by-single basis, weigh these socially-oriented discriminatory factors in the balance of whether or not to engage in genetic testing. Personal choices to test and selectively implant (or abort) do not negate our ideal of seeing everyone as created in God's image and valuing people on an individual basis for their achievements and their potentials, even if some disorders are being eliminated from the gene pool.

Today, we tamper with the "natural" trajectory of human lives to prevent or diminish suffering. For example, we prevent polio. For Spiritual Judaism, as many aspects of pain and suffering ought to be amenable to preventive human manipulation. Why then deny parents' desires that their children be born with a "sound" physical and mental constitution, not malformed or mentally deficient? Prenatal testing will spare parents the horror of watching children degenerate or suffer untimely death. It will help avoid marriages from being torn apart.

Spiritual Judaism suggests leaving it to each couple (or single parent) to assess whether they feel an embryo or fetus will result in a disabled child, subjecting him or her to physical or emotional suffering. They must consider whether a genetic defect will limit one's mental development (and to what extent) and even if pain can be controlled, what pleasure will be available to a child who lacks substantial physical mobility? Parents must also assess whether caring for a severely disabled child will be psychologically draining for them, as it generally. Not only may prenatal genetic screening allow the termination of a pregnancy if an embryo or a fetus has an unwanted genetic disorder but it may also permit appropriate preparation for a baby's birth, if the parents decide to go forward with a questionable condition.

Furthermore, the enhanced ability to test for and understand genetic causes of diseases may facilitate reproductive choice. Because of the increasing ability to test for genetic disorders, couples with a high risk of having a child with a genetic malady may be willing to have children. The availability of accurate tests on pre-implantation embryos will reduce the number of abortions and increase the number of wanted children.

Genetic Engineering: An Overview

Beyond genetic screening, physicians may some day use genetic therapy to cure conditions or diseases, through therapeutic change, inserting normal or modified genes to correct a genetic disorder, while a zygote is in vitro or a fetus is in the womb. Therapeutic genetic engineering will be used where the far safer and less complicated genetic embryo screening cannot be employed to deal with a range of diseases from the nearly inconsequential, to those impacting a person's quality of life, to the life-threatening by replacing the missing enzyme causing Tay-Sachs disease, or repairing the defective gene resulting in hemophilia or Huntington's disease. Someday, physicians will safely and effectively engage in preventive measures, curing a potential illness or condition in a fetus by rooting out a genetic defect or predisposition.

It is helpful to distinguish between somatic and germline intervention. Somatic therapy modifies or alters the body's non-sex somatic cells that do their thing and disappear when we die. Any genetic changes effect only the individual in question.[22]

With germline therapy, genetic changes are made in the reproductive cells that develop into sperm or eggs so that new configurations are passed on to future generations.[23] Perhaps sooner than we think, physicians will be able to tinker with the genes in sperm and eggs thereby changing future progeny. Allowing parents to take corrective action in their sex cells and thus eliminating genetic diseases or conditions from future generations is less expensive than somatic therapy in successive generations. It represents what experts have described as "the ultimate form of preventive medicine in that we can... prevent certain types of diseases even before a person is born."[24]

Therapeutic Genetic Engineering and the Jewish Tradition

The Jewish tradition, accepting and legitimizing therapeutic human intervention, extends to medical procedures prior to conception, in vitro and in the womb. Traditionalists and more liberal commentators approve of therapeutic genetic engineering. Gene therapy aims to prevent disease, restore health, and prolong life, all within a physician's Divine license to heal.

Gene therapy to replace the missing enzyme causing Tay-Sachs disease or repair the defective gene resulting in hemophilia or Huntington's disease is sanctioned under Jewish law. In these situations, physicians undertake gene therapy with the intent of restoring health and prolonging life. This type of gene therapy represents a legitimate implementation of the physicians' mandate to cure the sick, thereby doing the Divine work of healing and curing. Physician Fred Rosner, a traditionalist, maintains that therapeutic genetic engineering intervention serves as "a legitimate modification of the natural order," and does not "undermine God's creation of the world by manipulating nature."[25]

Liberal commentators take a similar, permissive view of gene therapy as an extension of current medical techniques and procedures. Because Jewish sages sought to avoid genetic problems, Conservative Rabbi David Golinkin reasons that they would approve of gene therapy aiming at eliminating serious or fatal genetic diseases both before and after birth. Any surgery permitted on an infant or an adult is permissible before birth.[26] Because the Jewish tradition has a presumption in favor of curing illness, Conservative Rabbi Elliot N. Dorff sees therapeutic genetic engineering as an "unmitigated" blessing.[27] One Reform responsum concludes: "We may be ready to accept genetic changes made for medical purposes and experimentation as [saving a life] is an overriding consideration."[28]

From a broader ethical perspective, The Council on Ethical and

Judicial Affairs of the American Medical Association has approved genetic therapy designed to avoid or treat genetic disease. The Council concluded, "Genetic manipulation in this instance [to prevent Tay-Sachs disease, Down Syndrome, or cystic fibrosis] would be a legitimate extension of the physician's preexisting and fundamental duty to combat disease."[29]

Thus, a consensus exists to allow humans use therapeutic genetic engineering to prevent diseases and disabilities. The next generation and those that follow would prefer health to serious disease or illness.

Therapeutic Genetic Engineering and The Extension of Life

Therapeutic genetic engineering, whether of the somatic or germline variety, offers the promise of extending the maximum human life span, doubling it to perhaps 150 to 200 years, by curing major genetically-based diseases, possibly tackling and arresting the aging process and curbing the indignity of bodily and/or mental deterioration. The longevity barrier, now about 120 years, may be torn down as we learn more about the genetic mechanism of cell aging and perfect intervention strategies. It may be possible to extend human life to perhaps 200 years, with the mind and body intact, preserving one's youthful appearance and strength, and warding off disease consistent with a relatively high quality of life and functional level until the end of one's life span. Individuals may have six careers, four master's degrees, three spouses. Women may have children in their 60s. Five or more generations may likely know each other.

The possibility that people will experience good physical and mental health for most of their old age will alter every phase of human existence. Certainly, more people will have several families with a higher proportion of stepchildren. However, given the already high incidence of divorce, it is uncertain whether marriages will survive for more than one century. Will other types of relationships come to exist to supplement marriage?

Making lives longer raises questions of profound social impact and presents the specter of titanic generational clashes, including economic consequences, specifically, how to fund Social Security and Medicare for the very elderly. Also, seniors may want to preserve the world as they know it. A ballooning in the population will increase the global demands for food, water, and energy further burdening the earth. Distributional questions will also arise if only the wealthy are able to afford these procedures that will begin with the embryo or fetus and continue thereafter. If only the rich can cheat death through expensive gene therapies, pressures will build to use public funds to pay for pre- and post-natal treatments.

Some diseases are linked to a single gene, such as cystic fibrosis or muscular dystrophy. However, most of the associations between genes and our health are much more complex, involving a combination of many genes often in hard-to-fathom interactions. Even if we master these complexities and engineer people with genes not only for longevity but also for reducing the susceptibility to illness, most diseases are multifactorial, that is, they result from a combination of genetic and environmental factors. Many ailments, such as lung cancer, arise from genes interacting with environmental influences. Unless environmental improvements are also undertaken, a 120 year life span may be a more realistic maximum age for most of us. Non-genetic and non-environmental factors, such as chance events, also play a role in good health and longevity.

Whether we live 80 or 150 years, questions remain as to how will each of us use our years. Will we lead different, better, more productive lives with more vital years to contribute? Will we be concerned with guiding many generations that follow us, passing on our acquired wisdom, values, and awareness? Despite our best efforts, earthly life will remain finite. Life will still gain value because we know we have a limited time and resources before we die. Our finiteness forces each of us to focus our time and energy on what is important in our earthly existence.

Non-Therapeutic Genetic Engineering

Genetic testing and therapy may serve non-therapeutic purposes, improving human appearance, functioning or performance, beyond what is needed to restore or sustain good physical and mental health.[30] Genetic testing may be used to screen and abort "defective" fetuses manifesting "undesirable" traits or characteristics, not diseases or conditions. Also, these traits or characteristics, such as height, hair or eye color, may be modified or changed by genetic engineering in vitro or in the womb. For instance, physicians will likely some day alter genes for height in pre-implantation embryos or fetuses. They may try to transform complex human traits, such as, intelligence or personality, each of which is coded by a large number of genes. By using these techniques, parents will be able to begin to design their own babies, eliminating genes for traits they regard as undesirable, adding genes for what they regard as desirable traits, and altering, to some degree, personality and intelligence. For purposes of the discussion that follows, let us assume that scientific research shows that genetic enhancement to be safe and effective.

Genetic Enhancement and the Jewish Tradition

The Jewish tradition remains divided on non-therapeutic gene manipulation, with the likelihood that most authorities will reach a negative conclusion. Nearly three decades ago, Azriel Rosenfeld, an Orthodox rabbi, supported the use of genetic techniques to enhance humans. Rabbi Rosenfeld[31] noted:

> Our sages recognize, and perhaps even encourage, the use of prenatal (or better, pre-conceptual) influences to improve one's offspring:
>
> Rabbi Yohanan (Babylonian Talmud Berakoth 20a, Baba Mezia 84a) used to go and sit at the gates of the place of immersion, a *mikvah* (the pool where women immersed themselves following their menstrual period so they could resume sexual intercourse with their husbands), saying: "When the daughters of Israel come out from their required immersion, they look at me that they may have sons who are as handsome as I and as accomplished in Torah as I."

It is unclear whether Rabbi Yonanan, whom Rabbi Rosenfeld referred to, intended only to give advice, not to promulgate a legal ruling. However, Rabbi Yohanan's admonition may well extend to allowing the use of genetic engineering to produce physically and mentally superior children.

Looked at from another angle, traditionalists may view genetic engineering on a egg or sperm before conception as permissible because genes are submicroscopic particles. Under Jewish law, any process invisible to the naked eye is not prohibited.[32]

Other traditionalists may be much more wary about whether non-therapeutic genetic engineering represents an appropriate use of technology. They may look to the Babylonian Talmud (Pesachim 54a) where we are told that after God inspired Adam with a type of Divine knowledge, Adam bred two dissimilar animals and created a mule. In response to the question, "Why are they called mules?" the Babylonian

Talmud answered, "Because they cast fear upon people" (Hullin 7b). By changing and denying the Divine creation of the world, Adam was viewed as inappropriately using nature. Thus, physician Fred Rosner, a traditionalist, asserts that non-therapeutic genetic engineering to improve physical traits and characteristics "is frowned upon in Judaism if it serves no useful medical or psychological purpose."[33]

Liberal commentators reach a similar negative conclusion. Conservative Rabbi David Golinkin would limit the use of gene therapy to the prevention and cure of diseases or conditions. He reasons that the Jewish tradition, permitting, if not mandating, that Jews be involved in medicine, speaks of using medical knowledge and procedures only for therapeutic purposes. Golinkin also bases his opposition on a repugnance for eugenics. He notes that societies using science and technology to improve human qualities historically have historically abused their power when pursuing eugenic goals. Golinkin also fears human-made "freaks" because there is no way of setting limits on smarter, taller, and stronger children. He concludes that non-therapeutic genetic engineering is "not in keeping with Jewish ethics and theology."[34]

Despite these misgivings and negative conclusions expressed by commentators, once we learn more about the genes for different temperaments, body builds, statures, and cognitive abilities, some prospective parents will likely want to choose some (or as many as possible) of their unborn child's traits and characteristics.

The Broader Implication of Non-Therapeutic Genetic Engineering

Beyond textual analysis and fears about eugenics, non-therapeutic genetic engineering, designed to pick or alter one's offspring's physical or behavioral traits, thereby fiddling with the genes controlling how we turn out, encounters strong resistance on a number of grounds, even assuming the requisite technical ability and the safety of the process. Opponents of non-therapeutic genetic therapy raise two major objections: first, it demeans human dignity and second, it presents distributional questions. They also offer two more minor reservations, namely, germline therapy gives too much power to the current generation and may increase the prejudice against disabled individuals.[35] Spiritual Judaism concludes, however, that none of these arguments stands up under scrutiny.

Genetic Enhancement Does Not Threaten Human Dignity

Focusing on the impact of genetic engineering on human dignity, critics raise two points. First, it impinges on the freedom of offspring and second, it turns them into commodities. These critics assert that children are autonomous beings, having their own lives to live. Subjecting a child's basic physical appearance and potential talents to parental whims impinges on one's ability to develop his or her unique self. A child's preferences concerning various benign traits or characteristics, they maintain, ought to be left to his or her values. In other words, decisions about benign traits or characteristics are not always best left to parents. Using genetic technology to determine what children will become, makes offspring partial hostages to parents' genetically designed blueprints.

Also, non-therapeutic genetic engineering may serve to treat embryos, fetuses and children, it is asserted, as commodities, objects of manufacture, raw material for manipulation, custom-ordered products, not as potentially sacred human beings. Allowing parents to manipulate a child's genetic characteristics to make him or her in the image of a couple's own preferences and desires may lead to a consumerist attitude to a child's development and personality. Use of these techniques may also dehumanize and depersonalize children undermining or significantly changing the meaning of the parent-child relationship, by reducing a child to an object of technological manipulation and design.

Spiritual Judaism recommends allowing parents to choose what they value in their offspring, whether to be disease free, at least genetically, or come with perks, such as musical talents or athletic ability. People today enhance their appearance or health through medical technology. The demand for cosmetic surgery shows how important people consider it to look good. Drugs are used to treat obesity. Short children receive growth hormones. It is not "immoral" to defy nature. We do it all the time.

Spiritual Judaism urges letting parents decide whether to take their chances on the traditional genetic lottery, using an unaltered egg and sperm, realizing that their offspring may inherit "undesirable" traits or characteristics, or undergo corrective gene changes. It recommends that parents permitted decide whether they want to turn to genetic modifications to improve the opportunities for their progeny.

As parents reflect on whether or not to use non-therapeutic genetic engineering, it is only natural that they want what is "best" for their children. Certain human abilities and characteristics are valued over others, these include height, intelligence, looks, an outgoing personality, and a stable temperament. Parents, desirous of providing their children with the greatest possible advantages will try to increase the odds that their progeny will "succeed." "Good" genes are presumably correlated with "success" in our society. Parents want their offspring to lead "better" lives. Some may want particular benefits, such as athletic ability or artistic and musical talents.

Social pressures will likely exist for parents to select certain traits or characteristics. Spiritual Judaism recognized that there are often social pressures to do many things throughout our lives.

Distributional Fears Raised By Opponents of Genetic Enhancement Are Overblown

Critics emphasize distribution concerns and socioeconomic disparities, focusing on a widening of the gap between the haves and have nots that may lead to the polarization of society into a genetically privileged ruling elite and a genetically disadvantaged underclass. People with money will be able to give their children not only a better environment, such as private schools, but also better genes. The wealthy will opt for children with the best set of genes -- emotional stability, long-term optimism and happiness, inborn talents, increased creativity, healthy bodies; while underclass families face obesity, heart disease, hypertension, alcoholism, mental illness, and a predisposition to cancer. Beyond the individual decisions of the wealthy, society may disparage children whose parents, affluent or not, allowed them to be born mentally dull, obese, or of short stature, among other imperfections.

If access to beneficial genetic technologies depends on income or wealth, the well-off will become healthier, at least overall. In the absence of the implementation of some system of universal healthcare in the United States or making these therapies an entitlement for all, these distributional concerns simply prove too much. In a modern capitalistic society, the wealthy and their offspring have (and will continue to have) access to opportunities, whether better healthcare or genetic services, that the economically disadvantaged do not have. Rather than leveling everyone by impeding the access to modern technologies, open the door to opportunity and encourage all to become more affluent through entrepreneurial activity, thrift, and sound, long-term investments.

Genetic enhancement technology likely will not remain expensive. Similar to the development of computers, where technological advances and the market economy made home computers cost effective and widely available, this type of technology will likely become sufficiently cheap for the non-wealthy. If not, the disadvantaged will probably not standby while biological alterations give some advantages over others, particularly a better chance to "succeed." Pressures will build for these advanced therapies to be an entitlement, with public funds providing access for everyone.

Children of well-off parents enjoy the advantages that wealth can confer. Parents try to improve their children's capacities and life prospects through private schooling, music lessons, and summer camp. Individual self-improvement is generally viewed as laudable. Engineering genes represents another way parents can strive to make their children "better" than the average in some way, and give them the best possible future.

Spiritual Judaism recommends that they be permitted to pass on these genetic gifts together with monetary benefits. It urges that we allow individuals to arrange for their offspring as they wish and permit them to provide not only the best environments for their children but also the best genetic natures.

The Benefits of Genetic Enhancement Outweigh Possible Discriminatory Impacts

Opponents point to genetic engineering as possibly exacerbating the prejudice against persons with disabilities or even those viewed as less than "perfect." Selecting offspring who conform "best" to social norms, for instance, health and physical ability, appearance, and aptitudes, may lessen society's tolerance for those with genetic or congenital disorders, whether shortness, less than a normal IQ, or a different physical build. The genetically superior may come to be contemptuous of their "inferiors."

Spiritual Judaism suggests that we let individuals weigh the possible impact of their personal choices on the social consequences, such as discrimination against the disabled, in deciding whether or not to avail themselves of modern therapeutic techniques. For Spiritual Judaism, the potential good outweighs the possible harm. Because we have devices to help the disabled cope, why not avoid the problems in the first place? Why would anyone want Tay-Sachs disease to continue, if we could stop it?

Germline Therapy Will Not Give Parents Control Over Subsequent Generations

The application of germline therapy in sperm and eggs raises the specter of the "dead hand" of parents exercising dominion over generations to follow. Again, fears about changing and potentially curtailing the autonomy of one's descendants for all time are overstated.

In all likelihood, germline therapies are not irrevocable and unalterable. Subsequent generations could elect to remove the enhancement, for instance, by undergoing the same type of procedure as installed the enhancement or in some way compensating for the effects of the enhancement.[36]

If it is objected that parents may make unwise choices, bringing grief to themselves and their progeny, that is indeed possible. However, these choices will likely be remediable; they probably will be able to be undone.

Even if some of these changes are irreversible, thereby curtailing the autonomy of and imposing on future generations a parental concept of health and values, Spiritual Judaism would allow the present generation to give future generations enhanced capabilities, for example, the capacity to be more intelligent, thereby lifting them to the peak of human potentialities. As an alternative to government regulation, Spiritual Judaism urges that we leave it up to each couple (or a single person) to make these choices in a responsible manner, on an individualized, decentralized basis. Through trial and error, let them use genetic technology to help their progeny. In exercising their autonomy, parents need to remember that what they choose may not turn out for the best in light of future conditions. Also, far from all of the basic characteristics of a human are genetically determined; our environment and our free will play roles. Because parents may be thwarted in attempting to control how their children's lives will turn out, they may come to accept the limits on their control.

Some Concluding Thoughts on Genetic Enhancement

A need exists to reject the rather simple-minded and naive biological determinism, ascribing everything to our genes, the qualities, personality, temperament, and intelligence, of who we are. The improvement of one's genes will not necessarily enhance a human's entire system in a beneficial and harmonious manner. Thus, it may be doubtful we will be able to bioengineer superathletes or superscholars.[37]

All children, including designer offspring, will still remain subject to the vicissitudes of their environment and chance events. Social and environmental factors have an incalculable and continuous influence on human growth and development, especially in the realm of personality and achievement; parents will come to realize they have some, but less than total control, over these forces. Thus, if parents give a child the athletic and musical genes, if such exist, to be a dance professional, for instance, the kid may turn out to be an uncoordinated klutz. Yes, genes confer an advantage, but we do not know (and may never know) what combination of factors yields a dance genius. Parents must remember that they may be frustrated when their designer children become persons other than what they had planned.

Accepting the reality that parents strive to improve a child's environment and that individual self-improvement is generally seen as laudable, even critics of genetic enhancement have softened their opposition, at least to some degree. For instance, the Council of Ethical and Judicial Affairs of the American Medical Association, concluded, "These considerations suggest that under certain circumstances, genetic manipulation [of benign traits or characteristics] could be justified as an extension of the continuing efforts of society, parents, and individuals to enhance the capacities and improve the lives of citizens."[38]

Finally, we should remember that the line between therapy and enhancement is blurry. Health is a continuum without a clearly defined stopping point. It is difficult to differentiate between treatment, prevention and enhancement. We lack an objective definition of "normal." If we see a correctable condition, most of us want to do something about it. Thus, interventions now regarded as enhancements may come to be viewed as therapeutic. If some day we correct the gene responsible for Alzheimer's disease, it may also be possible to enhance memory.

Parents contemplating the use of genetic engineering must ask themselves: what is "normal"? what is a disability (dwarfism)? what is a social problem (being short)? If genetic therapy gives us the capacity to change height and cure dwarfism, it seems unlikely that society will limit the therapy to a child to be in the first, fifth, or fiftieth percentile for height as an adult.

Spiritual Judaism recommends letting parents decide what they think is "best," what they regard as improvements of human traits or characteristics. Each of us, in light of social influences, has a different definition of "normal" health and desirable traits or characteristics. Many may want to design "perfect" offspring conforming to what they regard as societal norms, sociocultural priorities, or an economically "successful" genotype. However, these norms will vary thus limiting the imposition of harmful or skewed concepts of normality. Even if we follow a herd-like mentality, we likely will want to reduce aggression, stimulate friendliness, an raise intelligence, and eliminate genetic differences leading to inequality. Most of us would agree that at least these transformations do not represent the imposition of harmful concepts of "normalcy."

By giving people more control over procreation, genetic engineering does not represent an improper tampering with the "natural" order. On the contrary, it is itself an aspect of the "natural" order as human beings exercise our natural capacity to improve our lives.

In the past, a mother's and a father's genetic contributions were random. Once an embryo was created, its genetic character was fixed and we had no control over the expression of the genes and the development of the embryo. We were urged to accept fatalistically that this was the nature of things. We are now in a position to reject this attitude.

As co-creators and co-workers with God, in coming years we likely will have the opportunity to strive for excellence on behalf of our offspring, to eliminate genetic diseases, whether by genetic embryo selection or genetic therapy, and enhance our children, recognizing we are on the road to progress, not perfection. Spiritual Judaism suggest that we ought be able to free our children from the constraints of "natural" destiny. It supports the aspirations of people to live longer, to have better mental and physical health, and to give birth to children who have the best genes, the best chance for a long and healthy life, smarter and perhaps even happier.

Spiritual Judaism takes a dynamic, not a static, view of our nature as humans. We are not fated by God to accept the present state of things. We can change humans and their nature by technological means, providing a better life for our children.

Spiritual Judaism urges giving people as much choice as possible, provided these alternatives are safe and prudent. Thus, we ought to allow genes, through screening, therapy, and enhancements, to become instruments of personal growth and ever greater human freedom.

For some, the new biotechnology, resting on the premise that each human is just a pattern of information reduces humans to matter and extinguishes the sanctity of life. For Spiritual Judaism, the essence of human existence -- what is central to our humanity -- is our soul. Genetic intervention, the selection and design of children, does not contravene the preciousness of life. Its expression and development is intended to provide the biological basis for our sacred ensoulment, discussed in the next chapter. A good life involves wholeness, the uniting of body-mind-spirit.

The combination of genes with the environment and our individual souls is magnificent. This process contains so much mystery, deepening our belief in God, Who remains involved in the ongoing perfection of the world.

Notes

1. Dr. Immanuel Jakobovits, *Jewish Medical Ethics: A Comparative and Historical Study of the Jewish Religious Attitude to Medicine and Its Practice* (New York: Bloch, 1975), 266.
2. Meredith Wadman, "So You Want A Girl?", *Fortune* 143:4 (February 19, 2001): 174-176, 178, 180, 182, and Rick Weiss, "Virginia Clinic Develops a System for Choosing Sex of Babies," Washington Post, September 10, 1998, A1. E.F. Fugger et al, "Births of Normal Daughters After MicroSort Sperm Separation and Intrauterine Insemination, In-Vitro Fertilization, or Intracytoplasmic Sperm Injection," *Human Reproduction* 13:9 (September 1998): 2367-2370 discuss the successful use of flow sorting technology (flow cytometric separation), called MicroSort, to separate sperm cells for preconception gender selection. See also Sandra Ann Carson, "Sex Selection: The Ultimate in Family Planning," *Fertility and Sterility* 50:1 (July 1988): 16-19, at 17-18.
3. Fred Rosner, M.D., *Modern Medicine and Jewish Ethics*, 2nd edition (Hoboken, N.J.: KTAV, 1991), 123-131; Fred Rosner, "The Biblical and Talmudic Secret For Choosing One's Baby's Sex," *Israel Journal of Medical Sciences* 15:9 (September 1979): 784-787.

4. J. David Bleich, *Judaism and Healing: Halakhic Perspectives* (New York: KTAV, 1981), 111.
5. *Id.*, 111; Rosner, *Modern Medicine*, 123.
6. I have drawn on Richard V. Grazi and Joel B. Wolowelsky, "Preimplantation Sex Selection and Genetic Screening in Contemporary Jewish Law and Ethics," *Journal of Assisted Reproduction and Genetics* 9:4 (August 1992): 318-322, at 320; Richard V. Grazi, Joel B. Wolowelsky, Raphael Jewelewicz, "Assisted Reproduction in Contemporary Jewish Law and Ethics," *Gynecologic and Obstetric Investigation* 37:4 (May 1994): 217-225, at 221-222; Richard V. Grazi and Joel B. Wolowelsky, "New Ethical Issues," in *Be Fruitful and Multiply: Fertility Therapy and the Jewish Tradition*, ed. Richard V. Grazi (Jerusalem: Genesis Jerusalem Press, 1994), 186-189.
7. Bleich, *Judaism and Healing*, 112.
8. Central Conference of American Rabbis (CCAR), Responsum No. 160. "Predetermination of Sex" (1941), in *American Reform Responsa: Collected Responsa of CCAR 1889-1983*, ed. Walter Jacob (New York: CCAR, 1983), 509.
9. Daniel Schiff, "Developing *Halakhic* Attitudes to Sex Preselection," in *The Fetus and Fertility: Essays and Responsa*, ed. Walter Jacob and Moshe Zemer (Pittsburgh: Rodef Shalom Press, 1995), 102, 106. Schiff sees sperm separation is simple and safe. Also, a consenting married couple is involved in the process and a "legitimate objective" is sought. *Id.* at 102.
10. The Council on Ethical and Judicial Affairs, American Medical Association, "Ethical Issues Related to Prenatal Genetic Testing," *Archives of Family Medicine* 3:7 (July 1994): 633-642, at 636; The Ethics Committee of the American Society of Reproductive Medicine, "Sex Selection and Preimplantation Genetic Diagnosis," *Fertility and Sterility* 72:4 (October 1999): 595-598, at 597; Dorothy C. Wertz and John C. Fletcher, "Fatal Knowledge? Prenatal Diagnosis and Sex Selection," *Hastings Center Report* 19:3 (May/June 1989): 21-27; Dorothy C. Wertz and John C. Fletcher, "Sex Selection Through Prenatal Diagnosis: A Feminist Critique," in *Feminist Perspectives on Medical Ethics*, ed. Helen Bequaert Holmes and Laura M. Purdy (Bloomington, IN: Indiana University Press, 1992), 240-253.
11. Mary Anne Warren, "The Ethics of Sex Preselection," in *The Ethics of Reproductive Technology*, ed. Kenneth D. Alpern (New York: Oxford University Press, 1992), 232-245.
12. Wertz and Fletcher, "Fatal Knowledge?", 23.
13. Philip Kitcher, *The Lives To Come: The Genetic Revolution and Human Possibilities* (New York: Simon & Schuster, 1996), pp. 1-63, provides a useful introduction to DNA, genes, and genetic defects. See generally Matt Ridley, *Genome: The Autobiography of a Species in 23 Chapters* (New York: HarperCollins, 1999).
14. Nicholas Wade, "The Four-Letter Alphabet That Spells Life," *New York Times*, July 2, 2000, Section 4; Nicholas Wade, "Genetic Code of Human Life Is

Cracked By Scientists," *New York Times*, June 27, 2000, A1; Rick Weiss and Justin Gillis, "DNA-Mapping Milestone Heralded," *Washington Post*, June 27, 2000, A1; Sharon Begley, "Decoding the Human Body," *Newsweek* (April 10, 2000): 50-57; Natalie Angier, "A Pearl and a Hodgepodge: Human DNA," *New York Times*, June 27, 2000, A1.

15. Nicholas Wade, "Genome's Riddle: Few Genes, Much Complexity," New York Times, February 13, 2001, D1; Nicholas Wade, "Reports on Human Genome Challenge Long-Held Beliefs," *New York Times*, February 13, 2001, D1; Rick Weiss, "Life's Blueprint in Less Than an Inch," *Washington Post*, February 11, 2001, A1; Nicholas Wade, "Human Genome Appears More Complicated," *New York Times*, August 24, 2001, A13; Terrence Chea, "Tally of Human Genes Challenged," *Washington Post*, August 24, 2001, A10.

16. The successful use of preimplantation genetic diagnostic techniques to screen for and prevent Tay-Sachs disease is discussed in William Edward Gibbons et al, "Preimplantation Genetic Diagnosis for Tay-Sachs Disease: Successful Pregnancy After Pre-Embryo Biopsy and Gene Amplification By Polymerase Chain Reaction," *Fertility and Sterility* 63:4 (April 1995): 723-728. See also Alan H. Handyside and Joy D.A. Delanty, "Preimplantation Diagnosis: Strategies and Surprises," *Trends in Genetics* 13:7 (July 1997): 270-275 and Rick Weiss, "Building a New Child: Embryo Screening Creates a Tool Against Disease -- and Ethical Questions," *Washington Post*, June 30, 2001, A1.

17. Jakobovits, *Jewish Medical Ethics*, 155-156.

18. Fred Rosner, "Judaism, Genetic Screening And Genetic Therapy," *Mount Sinai Journal of Medicine* 65:5 and 6 (October/November 1998): 406-413, at 409.

19. Grazi and Wolowelsky, "Preimplantation Sex Selection and Genetic Screening," at 321-322, reprinted in Grazi and Wolowelsky, *Be Fruitful and Multiply*, 190-192.

20. J. David Bleich, *Contemporary Halakhic Problems* (New York: KTAV, 1977), 112. However, Rabbi Eliezer Waldenberg has ruled permissively on abortion for Tay-Sachs disease. *Id.*, at 113-115.

21. The Council on Ethical and Judicial Affairs, "Ethical Issues Related to Prenatal Genetic Testing," 637; David S. King, "Preimplantation Genetic Diagnosis and the 'New' Eugenics," *Journal of Medical Ethics* 25:2 (April 1, 1999): 176-182.

22. The scientific issues regarding somatic cell gene therapy are summarized in LeRoy Walters and Julie Gage Palmer, *The Ethics of Human Gene Therapy* (New York: Oxford, 1997), 17-36.

23. Scientific and technical background for germline therapy and its obstacles are provided in David B. Resnick, Holly B. Steinkraus, Pamela J. Langer, *Human Germline Gene Therapy: Scientific, Moral and Political Issues* (Austin, TX: R.G. Landes, 1999), 47-84 and Walters and Palmer, *The Ethics of Human Gene Therapy*, 60-76. See also Jane Lampman, "Doctors Cross Line That Modifies Future Generations," *Christian Science Monitor*, May 10, 2001, 1.

24. Resnick, Steinkraus, Langer, *Human Germline Gene Therapy*, 95.
25. Rosner, "Judaism, Genetic Screening and Genetic Therapy," at 408.
26. David Golinkin, "Does Jewish Law Permit Genetic Engineering on Humans?", *Moment* 19:4 (August 1994): 28-29, 67, at 29. See also David M. Feldman, *Birth Control in Jewish Law: Marital Relations, Contraception, and Abortion As Set Forth in the Classic Texts of Jewish Law* (Northvale, N.J.: Jason Aronson, 1998), 331.
27. Dorff, *Matters of Life and Death: A Jewish Approach to Modern Medical Ethics* (Philadelphia: Jewish Publication Society, 1998), 161.
28. CCAR Responsum No. 154. "Jewish Involvement in Genetic Engineering" (1989), in Walter Jacob, *Questions and Reform Jewish Answers: New American Reform Responsa* (New York: CCAR, 1992), p. 252.
29. The Council on Ethical and Judicial Affairs, "Ethical Issues Related to Prenatal Genetic Testing," 640.
30. The scientific issues pertaining to enhancement genetic engineering are summarized in Walters and Palmer, *The Ethics of Human Gene Therapy*, 99-108.
31. Azriel Rosenfeld, "Judaism and Gene Design," *Tradition* 13:2 (Fall 1972): 71-80, reprinted in *Jewish Bioethics*, ed. J. David Bleich and Fred Rosner (New York: Hebrew Publishing, 1978), 403.
32. Fred Rosner, "Genetic Engineering and Judaism" in *Jewish Bioethics*, 417.
33. Rosner, "Judaism, Genetic Screening and Genetic Therapy", at 411-412.
34. Golinkin, "Does Jewish Law Permit Genetic Engineering on Humans?", at 29. Rabbi David M. Feldman, "Forum," *Moment* 16:2 (June 1991):4, stated that the genetic enhancement technology is "interdicted for its own sake and for possible misuse." Earlier, Conservative Rabbi Lawrence Trester in "Therapy or Engineering: Jewish Responses to Genetic Research," *Reconstructionist* 49:6 (April-May 1984): 21-25, at 21, concluded that the production of a blue-eyed child as a preference "is unwarranted."
35. I have drawn on The Council on Ethical and Judicial Affairs, "Ethical Issues Related to Prenatal Genetic Testing," 636-638, 640-641; Mark S. Frankel and Audrey R. Chapman, *Human Inheritable Genetic Modifications: Assessing Scientific, Ethical, Religious, and Policy Issues* (Washington, DC: American Association for the Advancement of Science, 2000), 8, 31-32, 36-40; Resnick, Steinkraus, Langer, *Human Germline Gene Therapy*, 85-167; Walters and Palmer, *The Ethics of Human Gene Therapy*, 108-133; Glenn McGee, "Parenting in an Era of Genetic," *Hastings Center Report* 27:2 (March-April 1997): 16-22.
36. Walters and Palmer, *The Ethics of Human Gene Therapy*, pp. 85, 112.
37. Stephen Jay Gould, "Message From a Mouse; It Takes More than Genes to Make a Smart Rodent, or High-IQ Humans," *Time*, September 13, 1999, 62.
38. The Council on Ethical and Judicial Affairs, "Ethical Issues Related to Prenatal Genetic Testing," 640. The Council, however, considered genetic interventions to enhance traits as permissible only in severely restricted situations, satisfying three conditions. First, there needs to be a clear and meaningful benefit

to the fetus, or child who will be born, an important improvement, not the parental imposition of their own idiosyncratic values on their children. Second, there would need to be an absence of trade-offs with other characteristics or traits. Clear evidence must exist that the child would gain a benefit without a change in another characteristic or trait. Third, there would have to be equal access to the genetic technology, irrespective of income, wealth, or other socioeconomic characteristics, otherwise social divisions would widen rendering, rendering illusory the promise of equal opportunity for all. *Id.*, 640-641.

Chapter 9

Abortion, Selective Pregnancy Reduction, and Embryonic Stem Cell Research

Women get pregnant, men don't. Pregnancy and childbearing change women's lives forever, especially when faced with an unwanted pregnancy. Pregnancy and childbirth are not always positive experiences for women. They may be negative, even horrific. Unwanted children may live a miserable existence. Striving to preserve their own sense of self and to control their lives, leads many women to seek the right to "on demand" abortion.

Abortion is a serious matter. Spiritual Judaism supports a woman's right to choice and responsible autonomy. This is not to be equated with the irresponsible use of the right in the absence of carefully considered and sufficient reasons -- "pregnancy is going to ruin my vacation" or "having a child is inconvenient." Spiritual Judaism provides boundaries, teaching love and responsibility, in which exercise our autonomy and asserts that women (and men) think early and hard about aborting a pregnancy. No woman should cavalierly have an abortion; it should not be an act of selfish indulgence.

This chapter has three aspects. The first part of the chapter focuses on abortion, including maternal life-threatening situations and considerations of maternal pain, physical or mental anguish, as well as the fetus's future potential life and any disabilities it may face. For traditionalists, the fetus's right to be born is relative to and must be balanced against its mother's welfare. In the second part of the chapter, we turn to selective pregnancy reduction where multifetuses exist as a result of the use of assisted reproductive techniques or fertility drugs. Finally, the use of excess embryos for stem cell research is considered.

The twin principles of protecting potential life and viewing an unborn fetus as a "pursuer," not a person, limit traditionalists to sanctioning abortion only in limited circumstances, namely, a life-threatening or serious physical threat to the mother. The Jewish tradition does not accept abortion for the purpose of post-conception birth control. The life of the mother generally takes precedence over that of the fetus who does not have an independent existence separate from its mother prior to birth.

In seeking a middle ground between the clash of absolutes, namely, allowing abortion "on demand" for any reason at any time during pregnancy -- "it's a woman's right" and, at the other extreme, viewing the fetus from the moment of conception as a person -- "it's a baby" -- and thus any pregnancy termination constitutes murder, this chapter searches for a principled approach based on the concept of ensoulment conferring personhood. A need exists for women to have some degree of control over their reproductive lives if they are to fulfill their personalities, achieve a full life, and assume a position of equality in society. However, the fetus is not a nothing. Abortion represents the destruction of potential life and must not be undertaken lightly.

With choice comes responsibility -- first, to use safe contraceptive measures and second, not to abort beyond twenty-four weeks of gestation, except if pregnancy threatens a woman's life. Based on the most current scientific evidence, before the end of six months of pregnancy, a fetus's soul has not yet come into its body. Until that time it is too young to have a soul, even if it has a heartbeat, ten fingers, and ten toes.

Abortions in The United States

Since 1973, there have been well over one million abortions per year in the United States, many years reaching a staggering 1.5 million, at the same time one couple in seven or eight is infertile, many of whom are quite unhappy and desperate to adopt. In 1997, about 1.2 million legal abortions were reported in the United States. Approximately 30% of all pregnancies in the U.S. end in abortions, with somewhat over 300 legal abortions per 1000 live births.[1]

Despite these appalling numbers, the incidence of abortions in the United States has been falling for a number of reasons, including the aging of the population and a decline in unwanted pregnancies. Each year about 2% of U.S. women aged 15 to 44 now have abortions. Since 1980, when it was about 2.5%, the rate of abortions has declined.[2] Still, the United States remains the industrialized country with the world's highest abortion rate.

It is also useful to see at what point in pregnancies abortions are performed and the reasons women give for having an abortion. Approximately half of all abortions (53%) are performed before 8 weeks of gestation; approximately 86% are performed before 13 weeks. Few abortions are provided after 15 weeks of gestation, 4.2% at 16-20 weeks, and 1.4% at or after 21 weeks.[3]

As we will see, this last statistic, slightly over one percent, a minute fraction of all abortions, becomes a special focus of concern. Younger women, those aged 24 or under, are more likely to obtain abortions later in pregnancy than older women. The lateness in obtaining these abortions stems from their not realizing earlier that they were pregnant (or how long they had been pregnant) or having difficulty in arranging an abortion, for instance, teenagers who are afraid to tell their parents they are pregnant. Also, the increased use of prenatal diagnostic techniques, discussed in Chapter 8, raises the possibility of late term abortions occasioned by fetal abnormalities.

Women have abortions because of a variety of socioeconomic and family considerations. Many women seek abortions "on demand" because having a baby would interfere with work, school, or other responsibilities; they cannot afford a child; they do not want to be a single parent or are having relationship problems with their husband or partner. In one study, only three percent indicated that the most important reason for their decision was concern for their own health, with seven percent indicating this was one of their reasons. Three percent said that the most important reason was that the fetus had a possible health problem, with thirteen percent saying this was one reason.[4] Often, women have abortions because someone who loves them, a boyfriend, or parent, told them to terminate their pregnancy.

Abortion: The American Legal Background

Before developing the Jewish position on abortion, let's review briefly the American law on this topic. During the late eighteenth and early nineteenth centuries, following the pattern established by the English common (case) law, abortions were legal if performed before "quickening," the first recognizable movement of the fetus, typically from the sixteenth to eighteenth week of pregnancy. "Quickening," when a fetus moved and showed signs of life, represented the point at which potential life was to be protected. However, abortion was not a common law crime, even with respect to the destruction of a quick fetus.

A dramatic change occurred during the nineteenth century. By the turn of the twentieth century, legislation on the state level in the United States made abortion a crime at any stage after conception, unless done to save or preserve a mother's life.[5] States enacted anti-abortion legislation to protect women's health, to prevent the slaughter of lives, and to safeguard the privileges of the medical profession. In the second half of the nineteenth century, the newly formed American Medical Association (AMA) commenced a crusade against abortion. The AMA lobbied state legislatures for more restrictive laws against abortion as part of its campaign to bring sanctions against doctors who had not attended regular medical schools and were a competitive threat to those who had. The AMA also sought to crack down on errant "regular" physicians who performed abortions and thereby standardize the medical profession through disciplinary codes. Among feminists at that time, sexual abstinence, not abortion, was seen as resolving the problem of unwanted pregnancy. This consensus lasted some sixty years until about 1960.

Three factors fueled the trend to legalizing abortion in the United States.[6] First, pregnant women who took the drug thalidomide, a sedative, bore babies with deformities, such as flaps of skin instead of arms. As this link became known, women who had taken thalidomide and were still pregnant sought abortions.

Second, statistics concerning women who were injured or died as a result of illegal, back alley abortions reinforced the growing movement for less restrictive abortion laws. Policy makers saw the need to prevent women from dying from botched abortions.

Third, the feminist movement of the 1960s made the legalization of abortion, as a main component of women's sexual autonomy, a key priority. Women wanted the right to control their own lives. This led to a call for the legalization of abortion "on demand," abortion even without a therapeutic reason.

By 1973, about one third of the states, including New York and California, had liberalized their abortion laws, permitting abortion in certain circumstances, such as the pregnant woman's health, risk of abnormality or a serious defect in the fetus, or pregnancy resulting from rape or incest.[7] The United States Supreme Court in *Roe* v. *Wade*[8] accelerated the slow, ongoing process of state law revision.

Decided in 1973, *Roe* v. *Wade* rests on a pregnant woman's constitutional right of privacy, allowing her to make the decision to terminate pregnancy for herself, as it affects the use of her body, looking to her conscience. According to the majority opinion, government regulation of abortion served two important state interests, the protection of women's health and the potentiality of human life. The state's interest in regulating abortion for the sake of pregnant women's health could not be compelling during the first three months of pregnancy because abortions during this period caused fewer deaths than full-term pregnancies and eventual childbirth. During the first trimester, a woman's constitutional right of privacy allowed her to decide her future privately, without any inference from the state. Thus, women have an unconditional right to terminate pregnancy during this period.

After the first trimester, the majority in *Roe* v. *Wade* concluded, a state, "may regulate the abortion procedure to the extent that the regulation reasonably relates to the preservation and protection of maternal health."[9] During the second trimester, the right of the state to regulate and protect female health becomes compelling because of the dangers involved in abortion. Although a state cannot deny a woman the right to terminate her pregnancy, a state, based on its right to protect maternal health, may insist on reasonable standards of medical procedure should an abortion be attempted.

The state's interest in protecting potential life becomes "compelling" at the point of fetal viability, the period after which a fetus is potentially able to live outside the mother's womb albeit with artificial aid. The court placed this point at roughly from 24 to 28 weeks of pregnancy. States were given the power to prohibit abortion after the point of viability, "except when necessary to preserve the life or health of the mother." Viability was set as the "compelling" point "because the fetus then presumably has the capacity of meaningful life outside the mother's womb. State regulation protective of fetal life after viability thus has both logical and biological justifications."[10] During the third trimester, the state's interest in potential life overrides the pregnant woman's right to determine her own future for her own reasons without governmental interference.

Subsequently, in *Planned Parenthood of Southeastern Pennsylvania* v. *Casey*,[11] the High Court upheld its decision in *Roe* v. *Wade* but allowed a slightly looser standard for states to regulate abortion during the second trimester of pregnancy. Under the *Casey*, doctrine, when a state chooses to regulate abortion before fetal viability, it must ensure that no "undue burden" is placed on a woman seeking an abortion. The Court noted that some of *Roe's* factual assumptions had been overtaken by time, moving the point of viability from 28 to 23 or 24 weeks. These advances had "no bearing on the validity of *Roe's* central holding, that viability marks the earliest point" when the state's interest in the fetus becomes compelling.

More recently, a state ban on "partial birth abortions" was struck down by the Supreme Court in *Stenberg* v. *Carhart*.[12] The gruesome procedure in question, known as dilation and extraction, involves inducing labor to the point that the majority of the body comes out of the birth canal, while the head, still remaining inside, is punctured and its brain is suctioned out to facilitate the removal of a collapsed skull. Physicians use this procedure to terminate pregnancies after about sixteen weeks when the fetus's head has grown too big to pass through the undilated cervix.

The majority said the Nebraska statute in question failed to provide an exception to protect the health of the mother going beyond "situations where the pregnancy itself creates a threat to health" as required by *Roe* v. *Wade*. Note that standard geared to mother's health represents a lower medical threshold than maternal life. Second, the Nebraska statute was described as overly broad. The majority of the Court asserted that the open-ended prohibition could also jeopardize the most common method of second trimester abortions, dilation and evacuation, in which the fetus is dismembered and removed in pieces from the uterus, resulting in an arm or leg of a fetus often being pulled into a woman's vagina.

With this American legal background in mind, let us turn to a consideration of the relevant Jewish texts dealing with abortion.

Abortion: The Relevant Jewish Texts

For a text-oriented traditionalist, there are one biblical and several rabbinic sources dealing with abortion. These materials present two basic standards. The more restrictive focuses on the danger to a mother's life. The more liberal allows abortion in non-mortal cases, opening the possibility of abortions in a wider range of situations other than a threat to a mother's life.

In the Hebrew Bible, the Jewish view of the nature of the fetus is founded on a statement in the Book of Exodus (21:22-25). There, the author deals with a miscarriage caused by two men fighting and unintentionally pushing a pregnant woman. The text states: "If no catastrophe ensues, he shall be punished according to what the woman's husband shall exact from him; it shall be determined through adjudication. But if there is a catastrophe, then you shall give life for life."

Although it is unclear what the term "catastrophe" refers to, namely, the mother's or the fetus's death, most commentators interpret the term to mean the absence of fatal injury to the mother following her miscarriage. The person responsible for the miscarriage, the attacker, must pay money damages for the harm caused to the mother and the loss of the unborn fetus, but he would not be tried for murder as would be the case if the mother's life were lost. The assailant is not a murderer. It is no different from if he had unintentionally injured a woman elsewhere on her body.

An accidental abortion thus constituted a civil matter with a husband compensated for his loss of progeny. All of his wife's products, including the fetus, were regarded as his property. The accidental destruction of a fetus, an unintentional feticide, did not constitute murder, a capital crime.

Although not dealing with an intentional abortion, this biblical text assumes that fetal life has a lesser status than maternal life. Why? The text may reflect the author's view that an unborn fetus is not a person. This approach may lead to the conclusion that abortion is permissible, at least in some circumstances. Although the fetus is not a person, however, it still has some status.

Other sages maintain that the prohibition against abortion is rabbinic in origin. Relevant citations of rabbinic literature begin with Mishnah Ohalot (7:6) dealing directly with abortion in the context of mortal maternal danger. We are told that when "hard travail," that is, the life-threatening difficulty of childbirth endangers the life of a mother, an abortion may be performed and fetus extracted limb by limb. At certain times, even as late as the birth process, a mother's life, that which is most proximate to our reach, takes priority over a fetus's life, at least when the fetus is within her body. This approach leads to a permissive conclusion in cases of a therapeutic abortion to safeguard a pregnant woman's life.

The status of personhood is conferred only on a child who is "born." From the moment of birth, however, no interference with any natural process is permitted because "one life cannot be set aside for the sake of another life." The Jerusalem Talmud (Avodah Zarah 2:2, Sanhedrin 8:9, Shabbat 14:4) refers to "its greater part"; while the Babylonian Talmud

(Sanhedrin 72b) points to "its head" as signifying human life and personhood.

The noted medieval commentator, Rabbi Solomon ben Isaac, better known as Rashi, provided a standard for permitting abortion within the rubric of status.[13] Once a child is "born," he or she receives the status of personhood. He or she is then a full member of the human community. One person may not be killed to save another's life. As long as the fetus has not emerged into "the world," according to Rashi, it has not achieved the full status of personhood.

Using this line of reasoning, we can sacrifice an unborn fetus to save its mother's life. Abortion may be permissible in circumstances other than a threat to maternal life because of the inferior status of the unborn fetus. The unemerged fetus is considered part of its mother's body, not a separate being.

The Babylonian Talmud (Gittin 23b, Hullin 58a, Nazir 51a) describes a fetus as part of the mother, recognizing the dependence of the fetus on its mother and implying that the fetus has no, or at best a limited, status of its own.

This approach values the fetus only as a potential life, not as a person independent of its mother. Because any of her limbs may be amputated to save a woman's life, the fetus may also be destroyed for its mother's benefit. The act of birth changes the fetus's status. Liberal Jewish authorities and a minority of contemporary traditionalists, those of a more permissive bent, base their position on Rashi's logic.

Perhaps the crucial justification in the minds of many traditional commentators for limiting abortion to a restrictive set of circumstances comes out of discussion in the Babylonian Talmud (Sanhedrin 72b). The text encouraged the killing of a pursuer to prevent a potential murder of an intended victim. If one pursues another with the intent to murder, one must save the pursued even by taking the aggressor's life.

The concept of pursuit is based on two biblical passages. In the first, if two men are fighting and the wife of one pursues the other by maiming him, such conduct is not allowed (Deuteronomy 25:11-12). In the second, the author of the Book of Leviticus (19:16) tells us, "do not stand idly by the blood of your neighbor." Most commentators interpret this phrase to mean that one must not stand idly by without attempting to rescue another person whose life is threatened whether by robbers, drowning or wild animals. Another source for the law of the pursuer is found in Exodus 22:1. If a thief is found breaking into a house at night and is killed, the one who killed him, either in self-defense or by third-party intervention, is not guilty of murder.

Drawing on these biblical passages, the Babylonian Talmud (Sanhedrin 72b) recorded that:

> Rav Huna said: When a minor pursues another with the intent to kill he or she may be stopped, even at the cost of his or her life, in order to save his or her intended victim. Rav Huna reasoned that a pursuer needed not be forewarned as is the normal prerequisite to capital punishment, whether he or she is an adult or a minor. Rav Hisda objected: once the major part of a fetus has emerged it may not be touched, because the life of one person does not override the life of another person. Why is this case different? A difficult childbirth is not an instance of a pursuit as normally understood. The mother is being pursued from heaven, by an act of God. Her life is threatened by an unfortunate natural occurrence.

The Jerusalem Talmud (Avodah Zarah 2:2, Sanhedrin 8:9, Shabbat 14:4) had explained that in childbirth, we do not know who is killing whom or who is endangering whose life. Until its emergence from the womb, the fetus is viewed as the aggressor. Thereafter, we cannot make such a determination and the law of the pursuer does not apply; we may harm neither the mother nor the child to save the other.

This discussion in the Babylonian Talmud prompted Maimonides to categorize the fetus while still mostly in the womb and thus posing a threat to its mother's life as a pursuer, a stalker in modern parlance, coming to kill her.[14] Like any pursuer, the fetus, prior to its emergence from the womb, may be killed if it endangered its mother's life. Absent any threat to her life, an unborn fetus may not be aborted. The argument of pursuit becomes inappropriate once its head has emerged, marking birth. Then, if the child, who is now a person, endangers his or her mother's life, it is an act of God. It is a case of heavenly pursuit with the "natural order" pursuing the mother.

Maimonides' explanation remains controversial. We cannot automatically assume that Maimonides correctly applied the analogy of a pursuer to a fetus prior to birth. The Babylonian Talmud implies that no need exists for the theory of pursuer prior to actual birth. Rather, the Talmud reasons that the fetus's life is not as important as the mother's prior to birth, a position later advanced by Rashi. Also, Maimonides seems to use the pursuer analogy to justify a therapeutic abortion before birth, unlike the Talmud, if the pregnancy endangers a mother's life; yet he uses the Talmud's reasoning that once the forehead has emerged, infanticide is forbidden. The difficulty in understanding Maimonides and the proper explanation of his opinion remains of exceptional importance to traditionalists. His conclusion is paraphrased by *The Code of Jewish Law* as the relevant halachkic opinion on this topic.[15]

Authorities also look to the Babylonian Talmud's discussion of the execution of a expectant woman who had incurred the death penalty (Arakin 7a). According to the Mishnah, if an expectant woman were convicted of a capital crime and seemingly she did not disclose her pregnancy until judgment was decreed, she must be executed prior to giving birth. Judgments were to be promptly implemented. Only if she were in labor could the execution be delayed. Regarding this law, the Gemara comments:

> What is the reason [we wait if she has gone into labor]? Once the fetus has separated itself from her womb, it is considered to be an independent body [and no longer part of the mother]. R. Judah said that Samuel said: A [pregnant] woman who is on her way to be executed, they strike her against her womb in order that the fetus die first, so that [at the time of execution] she is not brought to disgrace [through the bleeding caused by the birth of the child].

Thus, the procedure that R. Judah described suggested the possibility of abortion in certain instances where the fetus did not threaten its mother's life, such as the possibility of a condemned woman's added shame. Some authorities use this text to justify non-therapeutic abortions for reasons apart from the mother's life or physical health.

Although showing concern for the mother's well-being as a significant factor, sometimes coming before the fetus's life, the text appears perverted in its attempt to protect the mother's honor.[16] Would not her humiliation be even greater if Samuel's procedure were followed? While alive she would experience the recommended procedure with considerable physical pain. Does avoiding possible humiliation at death justify causing great pain during life? Furthermore, who should make this decision? Should we not have asked the condemned woman what she would have preferred? Why should Samuel impose his concern for her disgrace at death, ignoring the woman's welfare during the fleeting moments of her life? Seemingly, if the pregnancy were known before the trial commenced or prior to the verdict being reached, would we not await the birth and then execute the woman?

In concluding that abortion is not homicide, other sages offer a different opinion as to the nature of the exact offense. We just discussed the rule that a pregnant mother sentenced to death must be executed so as to spare her the agony of waiting for her sentence to be implemented or the shame of miscarriage during the execution. The Babylonian Talmud (Arakin 7a)on this point states, "This is a simple ruling!" If abortion of the fetus were considered murder, the Talmud's statement "This is a simple ruling!" would be out of place.[17] The Talmud further states the mother is to be struck on the abdomen to cause the death of the fetus so as to avoid the desecration of her corpse resulting from the fetus' attempt to escape the womb after her death. If abortion were considered an act of homicide, the killing of the fetus would not be allowed to spare the humiliation of a condemned woman.

Two interpretations emerge from the biblical and rabbinic texts. The stricter version, following Maimonides, viewing the fetus as a pursuer and, thereby giving it the status of a human being, allows abortion only when a clear danger exists to the mother's life. The more lenient, following Rashi, taking the position that the fetus is subsidiary to the mother, seemingly allows other therapeutic reasons besides a threat to maternal life to be considered adequate to permit an abortion. Under either interpretation, however, the mother's welfare remains the primary consideration.

How Traditionalists Use These Texts Regarding Abortion

Among traditional authorities, a wide range of opinions exists as to the circumstances and the time when an abortion is permitted.[18] Apart from permitting an abortion within forty days of pregnancy, the consensus of traditionalists maintains the only acceptable circumstances for abortion are those in which the fetus poses a direct threat to its mother's life. Some traditionalists also consider dangers to maternal physical health a sufficient basis for abortion. Only a small minority permit abortion to safeguard a pregnant woman's mental health. Traditionalists nearly unanimously reject abortion for the sake of the fetus. Although stopping short of sanctioning abortion on demand, liberal authorities permit abortion for a wide variety of therapeutic purposes, such as maternal life, physical or mental health, or fetal defects, and some even going as far as recognizing socio-economic considerations.

Abortion Within the First Forty Days of Pregnancy

According to the Babylonian Talmud (Yevamot 69b, Niddah 30a-b), prior to the fortieth day of pregnancy a fetus is considered "mere fluid." At this early state of pregnancy, a fetus is hardly developed. Lacking form, it is not a human life and thus has no status at all.

Most traditionalists more readily allow abortions, even without any threat to a mother's life or health, during this early stage than during the remainder of a pregnancy. These authorities offer a lenient view, concluding that during the first forty days of embryotic development a fetus may be aborted for the slightest reason or perhaps even without any reason.

Others advance more stringent views.[19] They note that the Sabbath may be violated to save the life or preserve the health of an unborn fetus, even during the first forty-day period. The right to violate the Sabbath for the sake of a fetus, so that the child may later observe many Sabbaths (Babylonian Talmud Arakin 7b), demonstrates to these authorities that even during this early stage of pregnancy any abortion is proscribed; anticipating the future attainment of human status gives the fetus a sufficient status. Even within this analytical framework, these strict traditionalists would permit an abortion during the forty-day period if maternal life were threatened.

Abortion When Danger to Maternal Life Exists

Subsequent to the first forty days of pregnancy, when fetal formation begins to occur, the unborn fetus, although not considered a human life, has greater status. Yet, traditionalists permit, and in some cases, even require, an abortion to save a mother's life because the fetus is not considered a person until born.

In the absence of viable alternatives, abortion is required if the pregnancy poses a direct threat to a mother's life. If there is a mortal threat, that is, a clear and present danger, that maternal life would be at risk by continuing the pregnancy and non-lethal alternatives are not available, traditionalists maintain that Jewish law mandates that a woman have an abortion. We chose between the lesser of two evils. The mother is of this world; the fetus is not yet of the world. Commentators look to Mishnah Ohalot 7:6 that mandates performing an abortion for a woman in labor to save her life, provided the forehead (or a greater part) of the baby's body had not emerged from her womb. The mother's life takes precedence over that of an unborn child.

Following Maimonides's argument that once a fetus is torn loose from its mother's womb, the fetus is a pursuer who, similar to a murderer pursuing a victim, may be killed to save the life of the pursued. Thus, when the birthing process begins with the fetus detaching itself from its mother's uterus and the mother experiences a life-threatening birth, the fetus is seen as an "aggressor" pursuing its mother's life with the intent to kill her. The fetus' life must be terminated to save an innocent victim, namely, the mother. For text-oriented authorities, the Jewish tradition requires an abortion to preserve maternal life.

An abortion cannot be performed, however, once the baby's head (or a greater part of his or her body) emerges, even though the mother's life remains endangered. At that point, the newborn is considered a full-fledged human being. The baby's life is equal to the mother's. As we have seen, one life cannot be chosen over another. Because the danger to the mother then results from the Divine, specifically, a natural occurrence, the law of pursuit cannot be applied and the newborn cannot be sacrificed even for his or her mother's sake.

For the twenty-first century, we need to ask what is the relevance of the text (Mishnah Ohalot 7:6) to aborting a fetus during the early months of pregnancy, not during childbirth? Prior to the twentieth century, the danger to a woman's life in childbirth was extremely common. Also, with medical procedures that were dangerous or unreliable, early term abortions were uncommon, not so today.

Traditionalists permit an abortion, apart from childbirth, if carrying the fetus to term would pose a direct threat to the mother's life, a more than 50 percent probability that her life would be in danger if the pregnancy continued, for example, because of a risk of serious kidney malfunctioning. These authorities rely on sages in the Babylonian Talmud who speak of earlier stages of pregnancy where a fetus is considered as no more than a limb of its mother's body. In such a situation, it is not only permissible, but it may be mandatory, to perform an abortion to save maternal life.

A therapeutic abortion is required (or permitted) only if the danger to the mother's life results directly from the pregnancy. Sorting out questions of direct versus indirect causation represent difficult factual matters.[20] An abortion cannot be performed where doubt exists as to whether the pregnancy is the cause of maternal danger or it only complicates an existing condition. A pregnancy that only exacerbates an already dangerous medical condition threatening a mother's life cannot be terminated. The disease jeopardizes her life; the fetus, who only confounds a dangerous medical condition that would be the proximate cause of death, cannot be deemed an "aggressor."

Authorities who regard abortion as a form of murder refuse its sanction other than for the purpose of preserving a mother's life. Even if the fetus is the product of rape or incest, these traditionalists conclude that it cannot be destroyed. Others, relying on a different rationale, take a more permissive stance sanctioning abortion for maternal therapeutic reasons not involving a threat to her life.

Abortion When Danger to Maternal Physical or Mental Health Exists

Looking to the inferior status of fetus, more permissive traditionalists see menacing, but not life-threatening, situations, where a mother's superior interests may permit an abortion if the fetus poses a probable and substantial danger to its mother. Those permitting a therapeutic abortion where there is a physical hazard to the mother look to Rabbi Jacob Emden (1697-1776), who offered a radical and far-reaching ruling. Rabbi Emden allowed an abortion not only when the mother's physical existence was threatened, but also in a situation of "great need," for example, if continuation of the pregnancy would have caused her "excessive" pain and suffering. Under this standard, Rabbi Emden would permit an abortion prior to the beginning of labor if it served to save the mother from the "great pain" caused by the pregnancy; however, once the birth process had begun he prohibited abortion, unless the mother faced a life-threatening situation.[21]

Similarly, the late Rabbi Eliezer Waldenberg, the former head of the Rabbinical Court of Jerusalem, one of the twentieth century's preeminent Jewish bioethical authorities, permitted abortions if a pregnant mother's health were very precarious or she experienced severe pain.[22] Rabbi Waldenberg would permit the termination of the existence of a fetus with a conclusively proven fatal defect, such as Tay-Sachs disease, up to the end of the second trimester of gestation. He based his decision on mother's anguish at the prospect of bearing a fatally ill child. The abortion of such a fetus fell under the category of "great need" if its mother faced considerable mental pain in knowing her child would encounter a fatal outcome that might drive her to mental derangement, instability, insanity, or even suicide. Rabbi Waldenberg reasoned that "psychological suffering is in many ways greater than the suffering of the flesh."[23]

Only a distinct minority of traditionalists understand Emden's "great need" approach as incorporating both physical and psychological pain. This minority position would look to psychological as well as physical factors in evaluating the potential threat a fetus poses to its mother. These permissive authorities would allow an abortion to prevent emotional harm to a mother, specifically, a condition presenting grave consequences to her normal functioning. They develop a rather flexible line of reasoning, looking to the Babylonian Talmud (Arakin 7a), considering a mother's mental state or her worry or anguish caused by a child's potential deformity, among other problems. As discussed earlier in this chapter, the dignity of the condemned woman took precedence over the life of the fetus based on the mental anguish she would have experienced if she had to wait to complete her pregnancy before execution. For these authorities, emotional suffering provides a basis for abortion. The degree of mental illness or emotional upset, anguish, shame, or embarrassment that must exist, however, remains murky. In any event, the emotional criteria for an abortion, under this approach, require physician certification or a previous maternal mental condition, such as a post-partum depression.

In the late seventeenth century, Rabbi Israel Meir Mizrachi provided a ruling on the question of emotional health as a justification for pregnancy termination within a more traditional framework, concluding that serious danger to maternal mental health was the equivalent of risk to her physical well-being.[24] Because insanity created a distinct hazard to a mother's life he permitted an abortion when a pregnant woman was fearful she would suffer an attack of hysteria and become mentally deranged or suicidal. The noted twentieth century authority, Rabbi Moshe Feinstein, one of the more stringent authorities regarding the issue of abortion in general, concurred with Rabbi Mizrachi's conclusion, declaring that a pregnant woman's mental derangement may constitute a threat to her life.[25]

Most traditionalists do not, however, consider maternal psychological distress, even insanity, as a sufficient danger to maternal life. Reasoning that self-destructive inclinations are rare, they conclude that in most instances mental illness cannot be considered life-threatening.[26] Thus, those authorities prohibiting an abortion unless a woman faces a physically life-threatening situation would not permit the procedure for a woman who is experiencing or faces mental illness as a result of pregnancy.

A Liberal Perspective on Maternal Suffering

Both the Conservative and Reform Movements offer more permissive views on abortion, permitting abortions for a wide variety of reasons, including, serious danger to the life, physical or mental health of a mother, threats to the life or health of the child, as well as in cases of incest and rape. The Conservative Movement looks to the explanation of Rashi and the more permissive, traditional responsa as the basis for allowing the termination of pregnancy, provided serious maternal physical or emotional suffering exists. Reform rabbis also endorse maternal psychological considerations and some even go so far as to accept socioeconomic factors.

Conservative Rabbi Kassel Abelson, writing a majority opinion for the Conservative Movement's Committee on Jewish Law and Standards asserts that the true premise of the Babylonian Talmud (Sanhedrin 72b) comes closest to the explanation of Rashi, specifically, that the fetus is not considered a person until birth.[27] Disregarding Maimonides' explanation that the fetus is an "aggressor," Abelson concludes that Rashi's opinion provides the basis for the Conservative Movement's lenient rulings on abortion, while still providing a traditional Jewish source.

Conservative Rabbi Robert Gordis, in another majority opinion adopted by the Committee on Jewish Law and Standards, agrees with Rabbi Abelson, but takes Abelson's approach even further. In suggesting that Maimonides actually diverged from the Talmud, Gordis asserts that Rashi offers a fairly lenient, but correct, explanation on this topic. Gordis reasons that Maimonides felt it was necessary to preserve the Jewish people against extinction; therefore, Maimonides gave a more stringent ruling than required by the Mishnah and the Talmud. Gordis views the authentic interpretation of the text as resting with Rashi. Indeed, Rabbi Gordis sums up the Conservative view that abortion does not amount to murder, the killing of a person with a right to life. Also, he states that the best solution rests on "preserving a basically liberal attitude toward abortion with conservative safeguards."[28]

Within this permissive framework, Conservative authorities conclude that Jewish law requires abortion if pregnancy threatens a woman's life or health, either physical or mental. Also, abortion is permitted when the risk to a woman's life or health, physical or mental, is greater than normal pregnancy, even if the risk does not constitute a clear and present danger to her existence.[29]

Conservative rabbis, while permitting abortions for maternal therapeutic purposes, generally urge a liberal interpretation of the term "therapeutic," basing their opinion on lenient, traditional sources. For example, they would permit an abortion if the mother were a drug addict who caused the fetus's abnormality. In such a situation, the woman's worry represents a sufficient ground for abortion "because of the debilitating effects psychologically or otherwise on her well-being."[30]

For Conservative Rabbi David M. Feldman, a fetus can be sacrificed to save a mother's "spirit from torture and suffering." If a mother is worried that her fetus is deformed the fetus, a future potential, can be destroyed because "the mother is known, present, alive, and asking for compassion." Feldman would permit abortion of a defective fetus on the basis of preserving a mother's mental health if she were unable to deal with the prospect of bearing or raising such a child.[31]

The liberal approach of Conservative rabbis stops short of approving of a woman seeking to abort because she does not want another child. They conclude that abortion cannot be used for non-therapeutic reasons or as a form of post-conception birth control. A serious reason must exist to destroy a potential person, not merely because a pregnancy is unwanted or inconvenient. In so reasoning and reaching this conclusion, Conservative authorities reject the concept of abortion "on demand." The fetus represents a potential life that cannot be sacrificed in the absence of some reason.[32]

Within these general guidelines, Conservative authorities advise a pregnant woman to reach a decision in consultation with the father, members of her family, her physician, her rabbi, and anyone else who can assist her in assessing the "grave" issues involved.[33] Four points are noteworthy. First, a rabbi is not seen as the sole decision-maker. Second, halahkic sources should serve as one factor in a pregnant woman's decision. Third, abortion is viewed as a private matter, free from governmental regulation. Fourth, although abortion should be legally available, Conservative rabbis view it as the destruction of potential life and thus must not be undertaken "lightly or flippantly."[34]

The Reform Movement has consistently supported the right of a woman to obtain a legal abortion under the constitutional grounds the U.S. Supreme Court enunciated in *Roe* v. *Wade*.[35] Reform authorities have opposed interference in a pregnant woman's decision concerning abortion by any governmental unit.

Within the framework of its stance in opposition to governmental regulation of abortion, the Reform Movement generally offers a liberal view on the subject of abortion. Although agreeing with traditional authorities that abortion must be approached cautiously throughout a fetus's existence, one Reform responsum would permit an abortion at any time during pregnancy in case of serious danger, present or anticipated, physical or emotional, to the life of a mother (or the life or health of the fetus) as well as in cases of incest and rape.[36]

Another Reform responsum specifically permits an abortion for the sake of the mother's mental health. The question posed concerned a woman who contracted German measles during the third month of pregnancy. Because of the possibility that the child, if born, would be deformed in body or mind, the responsum concluded that for the mother's sake, her mental anguish now and in the future, she could abort the pregnancy.[37]

Some Reform commentators go far beyond maternal therapeutic grounds evidencing concern about a child who will not be properly cared for or one who will be born into a family where there are too many mouths to feed. For example, Reform Rabbi Balfour Brickner, focuses on the right of a woman, exercising her individual conscience, to make the decision. He finds it most desirable for women to choose how many children they feel can be adequately fed, clothed, and nurtured in a loving, familial environment. This freedom belongs to women alone; neither the state nor any religion can refuse them that right. Furthermore, Brickner does not understand why, in a world teeming with overpopulation and poverty, a woman ought not to be free to decide for herself whether to keep or abort her fetus.[38]

Brickner's approach is at odds with traditionalists who do not sanction abortion, except during the first forty days of pregnancy or if maternal life is threatened. Even those traditional authorities who allow abortion on maternal health grounds, as we have seen, represent a distinct minority.

Conservative Rabbi David M. Feldman stops short of giving women carte blanche. He concludes that abortion for population control purposes or economic reasons is not permissible. In a paper as a majority opinion by the Committee on Jewish Law and Standards, Feldman emphasizes the need to replenish the Jewish population decimated by the Holocaust.[39] The Conservative Movement's approach, while taking a more liberal stance than the traditionalists and permitting abortions for a wide variety of maternal therapeutic grounds, focuses on rebuilding the ranks of the Jewish people. For the Conservative Movement, the primary arguments against allowing abortions are not the Jewish legal sources or halakhic rules relating to it; rather it invokes a more stringent approach designed to rebuild the ranks of Jews wiped out in the Holocaust. Conservative authorities conclude that any pregnant Jewish woman considering abortion ought to seriously ponder the issue and the need to continue the Jewish people, who are not replacing themselves.[40] In other words, the Conservative Movement places the abortion decision within the context of a woman's responsibility to the Jewish people.

Brickner's view, by contrast, takes the opposite side of the spectrum from the traditionalists and the Conservative rabbis. He places the decision to abort in the hands of the mother by allowing her to choose whether she will be able to provide for her child. Looking to socioeconomic factors, according to Brickner, will promote the Jewish ideal of the sanctity of biological life. A fetus may be aborted if as a child it would not enjoy proper care. Brickner advances the idea that abortion ought to be encouraged, particularly in situations where social and economic hardships may impede offering the newborn adequate care.

Yet, limits exist to the Reform Movement's permissive stance regarding abortion. Stopping well short of allowing abortions for women who focus on their own convenience and comfort, one Reform responsum concludes, "We do not encourage abortion, nor favor it for trivial reasons, or sanction it 'on demand.'"[41]

In response to a question whether a woman may abort a fetus to relieve the potential suffering of other family members who would be burdened by caring for the child in the future, a Reform responsum returned to the longstanding focus of alleviating a mother's "great pain." Reasoning that "[f]etal life, though of lesser status than that of the mother, remains human life in potential, and is consequently of great significance...[to] be sacrificed only for the most profound reasons," Reform authorities would not sanction an abortion to spare other family members, apart from the mother, from anguish or hardship. Future worries regarding others do not constitute a threat to the mother's mental health, sufficient to justify the termination of the fetus' existence. A different situation would exist, however, if serious maternal anguish resulted from genuine fears with respect to a fetus' disability.[42]

Abortion of a Defective Fetus

Medical developments increasingly enable us to detect abnormal fetuses during the second trimester of pregnancy. Some defects can be detected by ultrasound screening that almost all pregnant women in the United States undergo. Physicians use ultrasound for women of all ages. Ultrasound, consisting of sound waves forming live video images of a fetus, enables a physician to see whether or not a fetus is maturing normally. Sonograms are also widely used as a means of prenatal diagnosis during the fifteenth or sixteenth week of pregnancy to diagnose a variety of fetal abnormalities, including anencephaly (the absence of all or a major part of the brain), hydrocephalus (a fatal swelling of the brain), spina bifida (a defective neural tube resulting in an exposed spinal cord that will cause paralysis of the lower limbs and impaired brain development). Sonograms, picturing a fetus in real time, make a fetus a reality for its mother.

Even more complex tests, such as amniocentesis, are routinely offered to pregnant women, especially those over the age of 35. Amniocentesis provides another means to determine fetal defects, including chromosomes leading to high levels of alpha fetoprotein indicative of congenital malformations, either anencephaly or spina bifida, or a genetic defect, such as Tay-Sachs disease. Physicians perform amniocentesis between sixteen and eighteen weeks of pregnancy when sufficient amniotic fluid is available.

Other prenatal tests are available. For instance, physicians use chorionic villi sampling to diagnose chromosomal abnormalities in a fetus between the ninth and twelfth weeks of pregnancy, far earlier than amniocentesis results can be learned. Three tests of maternal blood, the so-called triple screen, also detect certain genetic conditions, such as Down Syndrome. Finally, as discussed in Chapter 8, genetic testing will help give parents that a better indication of the likelihood that a child will experience an array of conditions and illnesses.

With the coming of sophisticated prenatal diagnosis making possible the determination of fetal defects, not only genetic but also metabolic, anatomical, and immunological, physicians will promptly respond to diagnosed abnormalities. Genetic testing will become a reality. Surgical therapies will exist for a wide range of conditions.

At present, invasive fetal surgery remains a risky procedure and is reserved for only the sickest of the sick, often resulting in premature delivery and increasing the chances of complications. Physicians currently use fetal surgery to correct a handful of rare lethal disorders, removing lung tumors, repairing diaphragmatic hernia (where organs slip through a hole in the fetal diaphragm, crowding the lungs and stunting their growth), cleaning urinary obstructions, decreasing handicaps caused by nonlethal neural tube defects, such as spina bifida where a child is born with an exposed spinal cord.

Apart from using fetal surgery or genetic therapy, if any of the various tests shows the fetus will be born with a "serious" physical or mental defect, for instance, a limited potential mental capacity or restricted ability to control one's own muscles, most women will opt to terminate their pregnancy. They are making a quality of life judgment, namely, a life with a "serious" disability is less desirable than life as a "normal" child. Most want a child to have a chance at least for an average, if not better, quality of life. Most regard it as foolhardy not to abort to prevent a life doomed to misery. It is not only what kind of life would the child have, but also the impact of its life on others' lives, its parents and its siblings.

Taking a dim view of this type of reasoning, text-oriented authorities maintain that an abnormal human being enjoys the same standing under the sanctity of biological life principle in Jewish law as a normal person. The same rules govern the abortion of a defective fetus as those regarding a "normal" fetus. Thus, if it is forbidden to abort a healthy fetus, apart from maternal considerations, this prohibition applies to an abnormal fetus, even one that is physically malformed, mentally deficient, or suffering from a fatal disease.[43] For nearly all traditionalists, only a mother's life or her health may be considered. Abortion for the sake of the fetus is not allowed. A pregnant woman's welfare provides the test for permissibility. The fetus is unknown and potential. If born defective, we do not know how bad its life will be. So much in the future we cannot foresee. There is the possibility of unanticipated forms of future treatment.

Among traditionalists, one notable exception was Rabbi Eliezer Waldenberg.[44] He would allow an abortion of a fetus, prior to the last trimester, if a definite risk existed that the fetus might be born physically or mentally handicapped, for instance, if it would suffer from Tay-Sachs disease.

A Liberal Approach to Aborting a Defective Fetus

Conservative and Reform authorities generally allow abortions for "defective" fetuses. Looking to a fetus's physical or mental makeup, they assert that a fetus with a "major" defect, precluding "normal" life, provides an independent basis for an abortion. For instance, the Conservative Movement has concluded that abortion is permissible "when the fetus is judged by competent medical opinion as severely defective."[45]

For these liberal authorities, the hard question focuses on what is a sufficient fetal defect to permit abortion? A fetus with a fatal, progressively degenerative disease, such as Tay-Sachs, or minimal brain tissue (anencephaly) are easy for Conservative Rabbi Elliot N. Dorff. Abort in such situations. But what about mental retardation (Down Syndrome) or a delayed degenerative disease, such as Huntington's disease that usually begins at age 35 or 40, suffers become depressed and violent toward themselves and others, may take their lives, slip into dementia, and lose bodily functions? Where the issue is clouded, Dorff advises judging the issue on the basis of a more traditional approach, namely, what is the mother's (or in today's egalitarian era, the parents') mental reaction to the defect.[46] Specifically, does she (or they) view it as manageable or beyond her (or their) emotional competence? Also, one Reform responsum would permit abortion in case of degenerative or permanent conditions seriously endangering the child's life and potentially the mother's mental health.[47] Dorff notes that only the most emotionally strong and stable people are able to raise "defective" children.[48]

Spiritual Judaism urges parents to undertake carefully their responsibility in deciding whether or not to abort a defective fetus.[49] They need to gather information about what life is like not only for children with the disability, but also for families with similarly disabled offspring. They ought to contact people living with the disorder in question. They also need to consider: the severity of the condition or illness; the age at onset; the probability of occurrence; if genetic, the probability that the genotype will manifest itself in a disease; the probability that the condition will occur in children or adults without the risk factor present in the fetus; and the harm to or burden for the family. For most of us, concern focuses on disabilities present at birth or those appearing in childhood, those that are more disruptive of what each of us conceives of as a "good" life. Most of us would likely abort for Down Syndrome, Tay-

Sachs disease, cystic fibrosis, or muscular dystrophy, but not Huntington's disease, an adult-onset disability, or asthma, an easily treatable condition.

The Search for a Reasoned Approach to Abortion: Science Helps Us Draw the Line

In view of the paucity and ambiguity of biblical and rabbinic texts dealing with voluntary, intentional abortions, particularly Maimonides' application of the pursuer analogy to a fetus prior to birth, modern commentators look to other rationales, such as what is acceptable or licentious behavior. More generally, how does permissibility or stringency regarding abortion comport with our idea of morality? How does the trend to the liberalization of abortion and the reasons for pregnancy termination fit into the moral compass of our society?

For Spiritual Judaism, the key to permitting or restricting abortion turns on the personhood of the fetus and its ensoulment. Although when the soul enters the body is said not to impact the Jewish tradition and its decisions regarding abortion,[50] in the twenty-first century we can no longer dodge hard questions, merely reciting ancient texts. Today, medical science enables us to better conceptualize what do we mean by the term "person?" when does human "life" begins? and at what stage does a fetus becomes a preborn human? Scientific evidence helps us start to answer other questions including: how should Judaism view the existence and the potential life of a fetus? how can we balance the fetus's existence and its potential life, an important consideration, against its mother's (and father's) wishes and needs? and what are the mother's (and father's) prerogatives with respect to the fetus?

Before looking at the soul in the Jewish mystical tradition and the modern evidence of ensoulment let us briefly go back to the ancient texts. The Babylonian Talmud offers various viewpoints on the subject of ensoulment. For some sages, an embryo is seen as endowed with a soul at the moment of conception (Sanhedrin 91b, 110b). In answer to the question (Sanhedrin 110b), "From when does a person acquire a place in the World to Come?", Rabinah stated, "From the moment that conception occurs, as it states, "Their seed shall serve the Eternal" (Psalm 22:31). Others place ensoulment at forty-one days after conception (Niddah 30a-b) or at the time of birth (Sanhedrin 72b, 110b; Genesis Rabbah Parashah 34:10), or circumcision (Sanhedrin 110b). Ensoulment may even take place as late as thirty days after a child's birth (Shabbat 135b; Yevamot 80b) or when a child first speaks (Sanhedrin 110b).

The Soul in the Jewish Mystical Tradition

The Jewish mystical tradition portrays the soul as a unified whole, one totality consisting of interrelated fields of awareness and energy connecting each individual to God. Jewish mystics view the soul, our unique inner substance, our essential spirit, our higher level of consciousness, as consisting of four intertwined energy fields or levels physical, emotional, intellectual, and spiritual.[51]

First, the physical level, a field of vital energy in each human being, attaches to, animates and preserves a human's body. It directs the functions of our organs and orders our vital processes. The physical level is sometimes described as a biological life force linking each human to the earthly world.

Second, the emotional level, a field of emotional energy, filling the heart with love, sympathy, tenderness and compassion, represents the locus of one's personality. This emotional level, sometimes characterized as emotional awareness, links the soul's physical level with its intellectual and spiritual aspects.

Third, the mental (or intellectual) level, a mental energy field as well as the energy of the transpersonal (nonphysical) self, connects each of us to the higher spiritual aspects of our existence. It is the invisible source of the mind's thoughts and inspirations, a channel through which our intellectual powers flow.

The mental level of awareness serves as a bridge between the human and the Divine realms. The development of an individual's mental level produces one's higher consciousness of God.

Fourth, each soul is characterized by a spiritual level consisting, in turn, of two spiritual or transcendental fields or levels. Within the soul's first spiritual level, each individual connects, consciously or unconsciously, with God and the Divine Life Force. Each of us experiences our separate self within the context of a larger whole. We strive to belong to this greater whole, to the Soul of the Universe, in other words, to God, and to find a context for our existence. We sense that somehow we can transcend our own existence, yet not lose our essence. The aspect of the soul represents the highest degree of awareness accessible to most humans.

But there is an even higher spiritual level, sometimes viewed as a transcendent field of light, in which the soul strives to unite and merge with God. This higher spiritual level constitutes the pure part of the soul.

The interaction of these four levels characterizes a human's soul nature, one's soulfulness. The soul represents the essence of a human being; it comes from and is a part of God.

From the perspective of Spiritual Judaism, we are interested in fetal physical and mental development, especially when ensoulment occurs, as a determinant of personhood.

The Scientific Evidence Regarding Ensoulment

Science helps us address when ensoulment occurs and personhood commences. Although science cannot provide us with conclusive empirical proof of when personhood begins, evidence of physical matters, such as brain structure and activity, helps delineate metaphysical issues, such as ensoulment. Once the fetus is a person, maternal considerations do not automatically assume priority over fetal life.

Fertilization, marking the sperm-egg interaction resulting in a single cell zygote, does not mark the beginning of personhood. A genetically and physically unique person does not exist at the point of conception. It is physically possible for more than one fetus to develop from a fertilized egg. Beginning about eight days after fertilization, spontaneous twinning may occur, producing more than one individual with identical genetic traits. Both twins have the same genetic blueprint as the original preembryo or zygote. About fourteen days after fertilization, one embryo cannot become two or more individuals through twinning. At that point, physical individuality is established. Twinning no longer occurs; twins do not recombine into one individual. Because a developmentally distinct embryo is not formed until the end of the second week after conception, it seems nearly impossible that ensoulment occurs on the fertilization of an egg.

Fourteen days after conception, the search for a biological basis for personhood can begin, focusing on fetal physical and mental development. In particular, a developing brain gradually lays down different systems at various times and begins to coordinate these systems.[52]

Zygotes, embryos, and early fetuses have no brain. They have no memory, no knowledge of their own existence, no intentions, plans, or goals for their future. They do not have a central nervous system. They have no mental life at all and lack consciousness. Fourteen days after conception marks the first appearance of a fetal nervous system.

At eight weeks, human features and all organs are present. The heart has been pumping for four weeks. The fetus responds to stimulation of its nose or mouth. The fetus has measurable, but precognitive, brain waves. However, the brain of a fetus is insufficiently developed for it to be sentient, to possess the capacity to feel pain or pleasure, to suffer or experience enjoyment. Spontaneous fetal movements occur. The fetal brainstem and spinal cord are present with neurons and synapses operational, but such early neural development does not constitute a brain-like organization.

The fetus needs sufficient brain structure to provide the basis for functional capabilities, reasoning, desiring, and relating to others, the attributes of human mentality. By ten weeks, inchoate, sporadic, disorganized electrical activity is detectable in the fetus's brain cells.

During the tenth to twentieth weeks post-conception, the construction of the crucial system of the brain commences. Electrical activity of the brainstem begins, but there is none in the cerebrum. No synaptic connections exist in the brain's cortex at this time. There is still no crucial system of the brain; thus no brain life occurs before the twentieth week.

Ensoulment and Personhood

When then is the fetus "alive" as a person in a meaningful sense? At what stage does its status as a preborn person come to the fore? Personhood depends on the presence of the brain's cerebral cortex, providing the physical basis for higher mental functions.[53] Before then, the fetus represents the potential for human life, but lacks the neural capacity requisite for human existence.

The fetal brain begins a rapid development between the nineteenth and thirtieth weeks. At twenty weeks, one part of the brain, its thalamus is connected to the cortex and the fetal nervous system is physically integrated. A functioning cortical brain does not occur prior to the twenty-second week.

At the twenty-second to twenty-fourth week, the fetus's cerebral cortex begins to mature, structurally and functionally. A functioning cortex starts to evince identifiable activity of the kind that normal adult cortexes indulge in, or the activity identifiable as a crude, undifferentiated ancestor of mature activity of this kind. With a minimal level of structural organization, evidence exists that the functional components are up and running and appropriately interconnected. The brain is structurally complete; all the important parts are there.

At this point, twenty-two or twenty-four weeks post-conception, the neocortex manifests minimally organized electrical activity with the commencement of electroencephalogram (EEG) waves. EEG activity exists, remaining discontinuous with bursts of activity for up to twenty seconds, followed by no activity for up to eight minutes.[54] Although the EEG activity is discontinuous, characteristic of an immature nervous system, it marks the beginning of cognitive activity and consciousness. Continuous EEG activity does not begin until thirty-two weeks of gestation.

Identifiable consciousness is in some sense correlated with some kind of activity in the cerebral cortex. There is electrical activity in the cortex that can be measured.

Consciousness represents a product of, or is in some sense identical with, some kind of activity in the cerebral cortex. There is electrical activity in the cortex which can be measured.

At about twenty-three weeks, a fetus's lungs become capable of respiration. Medical advances in the care of preemies make it possible to deliver a viable baby as early as twenty-three or twenty-four weeks. At this point, a fetus usually can survive outside its mother's womb. It possesses the capacity for a distinct human life beyond being part of a woman's body. It is capable of living independently. The basis exists for both the physical and mental aspects of life. The fetus is infused with a soul.

Spiritual Judaism and Abortion

The existence of a cerebral cortex marks the capacity for cognition and consciousness that correlates with ensoulment and personhood. At that point, the fetus has a structurally complete brain with all its important parts there. The brain has the capability to permit bodily functioning beyond spontaneous movement, the quickening marked by the earlier, perceptible, first felt fetal movements.

We connect the existence of higher brain activity with the possibility of cognition and consciousness and what is valuable about human life. A higher brain structure, in place and functioning, makes cognition and consciousness possible, even though these characteristics develop significantly after birth. A fetus with a cerebral cortex will ultimately be capable of problem-solving, rationality, and comprehending its environment. It will possess the ability to relate to others and form relationships that give satisfaction, recognize itself as a person, as a continuing subject of experiences and other mental states, and have a sense of personal identity. It will have the capacity to benefit from past experiences, to influence present situations, envisage a future for itself and have desires about its own future states, and be able to experience and express a range of emotions. It will eventually have thoughts and desires about the future.

At twenty-four weeks, a fetus possesses the minimal, but crucial, level of structural brain organization. The functional components of its brain are present and mature enough to perform some of the identifiable activities -- the very primitive cognitive and consciousness functions -- of the kinds characteristic of normal adult brains. Thus, it seems likely that ensoulment occurs with the existence of the brain's cerebral cortex.

On the basis of our scientific knowledge, Spiritual Judaism would allow abortions through the twenty-four weeks of nonviability and nonensoulment, provided there is a sufficient justification involving either maternal life or health or fetal health. After six months of gestation, the fetus is more than merely responding to pain, making respiratory efforts, exhibiting a heart function, moving spontaneously. Its brain waves are discernable by means of an EEG. As one Reform responsum concludes, "We must, therefore, be more certain of our grounds for abortion, but would still permit it [until the fetus becomes viable if there is a serious danger to the child's or mother's health]."[55]

Throughout a woman's pregnancy she must seriously consider the fetus. Spiritual Judaism recommends against a woman terminating her pregnancy at whatever time and for any reason she chooses. Prior to the end of the second trimester of gestation, the claims of a pregnant woman, are, on balance, stronger than those of a fetus. Although the fetus is not a person, it still has some status. It should not be treated as a nothing or destroyed without good reason. Sufficient reasons include maternal life or health considerations (physical or emotional) or the existence of deformed or seriously defective fetus. Ending a fetus's existence before it is sentient, before it feels pain, provided a sufficient reason exists, respects, in the view of Spiritual Judaism, the value of human life.

In addition to discerning the existence of a sufficient reason to abort, it is important to ask from whose perspective the decision ought to be made, the mother alone or mother and father? In other words, who should enter into the decision-making process?

Although the Jewish tradition looks at abortion from a maternal perspective, Spiritual Judaism recommends considering the relationship of a fetus to both its mother and father. During pregnancy, both its mother and a father generally develop a significant relationship with the fetus, even before its birth. Especially in our age of sonograms, a fetus has a reality, it has a presence. The sonogram's ultrasound image serves as a baby's first photo, facilitating the bonding of parents and the child-to-be. Spiritual Judaism recommends, therefore, involving the father in the decision-making process, though leaving the final decision to the mother.

After viability, Spiritual Judaism would counsel against abortion, except to save a mother's life or prevent a "serious" threat to maternal life, but no longer merely because her physical or mental health. Even at this stage, the maternal life takes precedence over fetal existence. Once ensoulment occurs and the fetus becomes a "person," its status overrides any defect in the fetus or a woman's therapeutic considerations, apart from a "serious" threat to her life.

A Preventive Approach

Beyond the difficult questions relating to abortion we have been considering, perhaps the real problem lies in the epidemic rate of unintended pregnancies in the United States. Some fifty-percent of all pregnancies in the United States come at an inopportune time or are unwanted altogether.[56]

With the current, ready availability of abortion as a handy post-conception method of birth control in the United States, Spiritual Judaism advocates reasserting the virtue of responsibility, a rather strange notion in our era of irresponsibility. We need to deal with the problem in the first place by prohibiting casual pre-marital sex, as discussed in Chapter 3, and through the use of effective contraceptive methods. Sexually active men and women who do not want to conceive have a responsibility to practice contraception; they must protect themselves against unintended pregnancies. Women need to choose childbearing when it is most appropriate for them; for most women, when they are married and have a husband who earns enough money to support a family. In other words, make childbearing intentional, at a stage in a wife and her husband's lives when they can raise a child with the optimum conditions for his or her well-being. The more thought given to the choice to conceive or not, the more likely a carefully considered decision will eliminate, or at least lessen, the need for abortions, apart, of course, from maternal or fetal health considerations.

Our ability to limit the risk of unchosen pregnancy before conception turns on the use of effective birth control techniques. Barriers to birth control and information about it continue to exist. For some, it is cost or even availability; for others, poor sex education, religious proscription or cultural tradition. Apart from need to remove the informational barriers to contraception for those who want it, emotional factors enter into the equation. Is it possible to have a sense of responsibility and the accompanying discipline in the heat of passion? Women, especially unmarried females, are expected to provide spontaneous, fulfilling sex, as they are swept away with passionate surprise. Some women do not use birth control measures because they do not want to appear to their male partners as though they expected sex or they want to force a man to take their relationship more seriously.

Senseless mistakes continue to occur. Couples are careless about preventing conceptions they do not want; some of us have experienced irresponsible sex. A woman's life can change radically, even crumble quickly, when faced with an unwanted pregnancy. We cannot downplay the human suffering of compelling women, often those whose who are poor, ignorant and frightened, to bear children they do not want and therefore may resent. Pregnancy and motherhood represent a loss of control of many women's present and future selves. Neglect and abuse can result for many children from growing up unwanted.

In the ideal world, abortion would remain a rare option to be used in very limited situations. Only wanted children would be brought into the world.

In moving from the present to the ideal, Spiritual Judaism urges that we permit, but discourage, abortion and support adoption. A clear need exists to replenish the Jewish population in the twenty-first century even in the context of serious concerns about planetary overpopulation.

Spiritual Judaism recommends focusing on pregnancy prevention, making abortion less necessary. It bears repeating -- do not indulge in unprotected sex; intercourse can lead to conception. The non-use or misuse of contraceptives may play an important role in the occurrence of unintended pregnancies. About half (47%) of all unintended pregnancies occur in women who do not use contraceptives at all. A majority (53%) of the remaining unintended pregnancies result from the inconsistent or incorrect use of contraceptive methods.[57]

Spiritual Judaism advocates preventing unwanted pregnancies through responsible sexual behavior, specifically the use of male and female contraceptives. Both men and women must take the responsibility for better birth control practices, consistently and correctly used. Couples must practice contraception more effectively, changing from less efficient methods to those with lower failure rates.

Let us hope that the future will see the development of more effective contraceptives, easier to use, with fewer side effects. We may also witness greater use of post-intercourse contraception techniques, for example, the morning after pill to negate contraceptive failure, such as the breaking of a condom, and unprotected sex or other careless mistakes.

A detailed discussion of contraception in the Jewish tradition is beyond the scope of this work.[58] Suffice it to say that traditionalists treat male contraception, specifically, the use of condoms, far less favorably than female forms of birth control.[59] Concern over the male duty to procreate and violation of the prohibition against the wasting of seed, discussed in Chapters 3 and 4, exists. Some of these text-oriented authorities prohibit the use of any contraceptive by males. For them, the normative ideal remains a large family and a lifelong commitment to procreation, relying on God's providential care. Contraceptives for females have gained greater approval, especially oral contraceptives and intrauterine devices, at least those used on medical grounds to safeguard life or health.

The liberal strands of Judaism, especially the Reform Movement,[60] support the use of a wide variety of birth control techniques. However, the Conservative Movement maintains that Jewish law argues against a man using a condom at least until he has fulfilled his duty to procreate. For women, Conservative authorities recommend the use of a diaphragm because it has basically no impact on a woman's health; a contraceptive pill or implant is the next most favored means of female contraception, if compatible with a woman's age and body chemistry.[61]

Reproductive responsibility also means taking steps to reduce the possibility of multiple fetuses thereby avoiding the need for selective pregnancy reduction, to which we now turn. With in vitro fertilization, physicians typically transfer an average of four embryos to a woman's uterus hoping at least one will successfully gestate. It is better to implant only two or three embryos conceived in vitro outside a woman's body.[62]

Selective Pregnancy Reduction

In a physician's office, the shadowing images flicker on an ultrasound screen, defining four viable embryos in a womb of a previously infertile woman. In vitro fertilization, an assisted reproductive technique discussed in Chapter 6, was successful, perhaps too successful. When her physician transferred six of her fertilized eggs to her uterus, it was in the hope that one or two would attach and grow. Now, she is pregnant with multiple fetuses.

For a couple who cannot conceive, modern reproductive technologies balance hope against the percentages. The more fertilized eggs that go into a woman, the better the chances of at least one embryo developing. Typically, three embryos result in one child; four in one or two. But sometimes four embryos can become four fetuses.

Some women are given strong doses of hormones designed to stimulate their quiescent ovaries to release eggs. Fertility drugs that stimulate ovaries to produce many eggs at once are available by prescription through willing physicians. About 10% to 20% (or more) of previously infertile women in whom ovulation is induced become pregnant with multiple fetuses.[63] Thus, in vitro fertilization or infertility treatments may lead to multiple births, supertwins, that is, triplets or more (quads, quints, or more).

Women are not designed to have supertwins. Most can safely carry up to three fetuses; however, more than three fetuses pose a substantial risk to the health and the life of the mother and the fetuses. The risks are directly proportional to the number of fetuses mainly because of intrauterine growth retardation and increased likelihood of premature delivery and the attendant problems, including critically low birth weight, and lung and intestinal infection and bleeding. One in ten infants from supertwin gestations dies before reaching twelve months.[64] The survivors of multiple gestations often face a long-term outlook marked by congenital malformations, such as anencephaly, blindness, or cerebral palsy as well as developmental, physical, and learning disabilities.[65] The mothers of multiple fetuses are also subject to extensive pregnancy complications, including high blood pressure and diabetes.

To lessen or eliminate the maternal and fetal complications associated with multiple fetal pregnancies, specifically, the high fetal morbidity and mortality rates and maternal morbidity, physicians often recommend reducing the number of fetuses in a woman's uterus so that those remaining have an improved chance of surviving normally.

A woman pregnant with supertwins and her husband face a difficult, if not heart wrenching, choice. Should she undergo selective abortion, the purposeful abortion of some fetuses to make it possible for one or two to live, or permit a multiple fetus pregnancy to proceed, possibly endangering her life and the development of the unborn fetuses?

Selective pregnancy reduction eliminates some fetuses before they are born. It is typically performed before the end of the third month of pregnancy by injecting selected fetuses with potassium chloride that stops the heart. A physician, guided by ultrasound, inserts a needle through the abdomen or vagina of the mother-to-be to deliver the lethal injection that kills some of the fetuses.

In formulating an approach to selective pregnancy reduction, the Jewish tradition must wrestle with a number of complex questions. May one sacrifice some fetuses so that others may be born healthy? What about the selective abortion of one or more fetuses having a serious physical or mental defect or a genetic deformity so as to allow the other normal fetus (or fetuses) to be delivered?

Pregnancy Reduction Within Forty Days of Conception

It is considerably easier for traditionalists to permit selective reduction within the first forty days of conception. As discussed earlier in this chapter, many text-oriented authorities permit abortion prior to forty days after conception, even in the absence of any threat to a mother's life or health. Nothing yet has happened. There is no harm to any fetus. Reasoning that "the small embryo is not considered to have a firm claim on life. This is especially so when each embryo is endangered by the presence of others," traditionalist physician Fred Rosner and Orthodox Rabbi Moshe Tendler conclude, "[I]t seems likely that they [most Orthodox rabbis] will be lenient and allow the destruction of some pre-forty-day-old embryos in order to allow the others to mature and be born healthy."[66] This position seems valid especially where none of the fetuses will survive if the pregnancy proceeds without intervention.

Threats to Maternal Life

As we have seen earlier in this chapter, most traditionalists permit selective abortions if a threat exists to maternal life. When the implanted fetuses threaten her continued existence, each fetus is regarded as an "aggressor." Because a mother's life takes precedence over that of the aggressor, any fetus may be sacrificed to save her life.[67]

Threats to Maternal Health

When the danger to maternal health is not life-threatening, as previously discussed, disputes exist among traditionalists. Some, such as Rabbi Eliezer Waldenberg, permitted selective pregnancy reduction in this situation.[68] Fred Rosner maintains that because multiple pregnancies pose a high rate of serious maternal complications, it may be permissible to destroy one or more fetuses to reduce or eliminate serious maternal health risks.[69] Those traditionalists, who maintain that abortion represents a form of homicide, would not, however, allow pregnancy reduction unless the mother's life were at stake.

Liberal authorities are more permissive. For example, selective pregnancy reduction to protect a woman against a serious threat to her health is acceptable to the Conservative Movement.[70]

Dangers to Fetal "Lives"

In the absence of any danger to maternal life or health, under normal circumstances, nearly all traditionalists prohibit abortion. However, this new topic gives rise to a unique question in halacha, namely, whether it is permissible to abort some of the fetuses in a situation where all are doomed unless such a procedure is performed, in order to save the others, applying the theory of "aggressor" to the aborted fetuses. The complexity of this issue is compounded by the fact that all of the fetuses are, in essence, considered aggressors in relation to the others. The fetuses are equal in status. Each fetus is an aggressor as well as a victim. We cannot single out one or more fetuses and declare it (or them) to be the pursuer(s). Thus, it may not be permissible to kill one for the sake of another. Furthermore, members of a group cannot select one of their group and deliver him or her to death in order to save themselves because each individual's biological life has value.

In the context of selective pregnancy reduction, each of the multiple fetuses is equal. Each has its own claim to continued existence. No single one has been designated for death. Despite the likelihood that all will die if nothing is done, many traditionalists would likely forbid the sacrifice of one or more fetuses to save the rest, again absent a threat to maternal life (or perhaps health).

Where a danger exists to all fetal lives and none will likely survive if nothing is done, some traditionalists may permit selective reduction provided it is done on a random basis.[71] They could base their conclusion on the reasoning advanced by the Babylonian Talmud (Shabbat 135b; Yevamot 80b), namely, a newborn who cannot survive for thirty days is not considered a living human being. Because all of the fetuses are doomed to death (and are categorized as non-viable) unless some are aborted, no prohibition exists against aborting some. However, if any of the fetuses could survive, even though as newborns they may likely suffer some defect or abnormality, even these more permissive traditionalists likely would not allow the procedure.[72]

Modern Orthodox Rabbi J. David Bleich would permit selective pregnancy reduction to save the remaining fetuses who need the intervention to establish eventual viability. A fetus who cannot survive, Bleich asserts, is never considered a human life, even in utero; its diminished status was determined from conception. Pregnancy reduction involves the destruction of some fetuses to save the remaining ones, who will become viable. Bleich maintains that without selective reduction, none of the fetuses would survive. Thus, this is the only form of rescue available for the other fetuses. Those eliminated were not endowed with viability; therefore, their elimination, Bleich maintains, is permitted.[73]

Liberal authorities evidence a permissive approach in situations involving danger to fetuses, not their mother. For instance, Conservative authorities permit selective reduction of multiple fetuses under the general Jewish legal principle of saving lives, here, protecting the remaining fetuses.[74]

Dangers to Fetal Health

Traditionalists debate the efficacy of permitting selective pregnancy reduction in the absence of a determination that without this procedure all the fetuses will die. As we have seen, traditionalists overwhelmingly do not sanction abortion for the sake of the fetus. However, in this situation, permissive traditionalists reason that because the greatest number of rabbinic opinions allow the destruction of one or more nonviable fetuses, they would allow a physician to abort one or more fetuses facing a threat of physical or mental abnormality. These authorities urge that it is better to do this as early as possible, particularly within forty days of conception, as this is the period within which most traditional authorities permit abortion.[75]

The Conservative view mirrors the lenient traditional position. According to Rabbi Elliot N. Dorff, when it is possible, for example, through genetic testing or other means, to determine the number of viable fetuses, it is permissible to abort selectively those with less of a chance to survive and be healthy.[76] This view comports with those opinions allowing pregnancy reduction even if the fetuses are only threatened with developmental abnormality.

Selecting Fetuses to Abort

If a selective abortion is permitted, it is up to the physician involved to decide the number of fetuses to be destroyed to ensure a good prognosis for the mother and those remaining. If all of the fetuses are equally viable, physicians usually opt to take the ones that are the most accessible, at the time of the procedure. If all are equally accessible, they abort on a random basis.[77]

Excess Embryos and Stem Cell Research

The in vitro fertilization procedure, discussed in Chapter 6, often produces more embryos than are required for a pregnancy. Some of these excess embryos may be frozen and stored.

An embryo put into the deep freeze typically consists of about one or two hundred cells, each of these cells containing all the information needed to start a unique genetic existence. Researchers take stem cells from a frozen embryo, four to seven days after fertilization, when it is a hollow sphere consisting of one or two hundred cells. Inside the sphere is an inner cell mass containing stem cells.

These embryonic stem cells offer vast medical potentials. The primordial, undifferentiated mass of cells, not yet specialized into brain cells, heart cells, or other kinds of cells have remarkable ability to develop into nearly any of the body's specialized cells and tissues, including those of the brain, heart, and liver.

Potentially representing one of the biggest breakthroughs in modern medicine, stem cells may someday be used to rehabilitate tissues and organs damaged by illness or accident or restore immune systems.[78] As a focal point of regenerative medicine, they may be used to treat degenerative diseases and generate an unlimited supply of replacement tissues and organs, to grow new heart muscles for heart attack survivors, insulin secreting cells for diabetics, and neurons for Parkinson's disease sufferers. Such replacement parts could extend human life spans by decades, with a significantly enhanced quality of life. At present, stem cells for research purposes are taken from surplus embryos donated by couples (or single mothers) to fertility clinics. Furthermore, many of the body's tissues possess a store of adult stem cells that may be able to perform some of the tasks expected of embryonic stem cells.

Researchers are currently striving to perfect and develop stem cell techniques. These cells have the capacity to grow and divide indefinitely in a test tube if fed the proper mix of chemicals. Stem cells give rise to offspring cells that have the ability to reproduce and under the right influences to specialize into nearly all of the body's cell types, regenerating Parkinsonian brains, rotting livers, or decaying hips.

Scientific uncertainties exist. It is, at present, difficult to keep stem cells alive in the laboratory. Furthermore, it is nearly impossible to grow them in numbers large enough to be medically useful, hard to cultivate them into the specific kind of cell physicians want, and keep them from specializing until researchers and clinicians are ready. There is an additional problem. We do not know how to engineer the stem cells so that they will not face rejection.

Experimentation with embryonic stem cells, obtained from spare human embryos, lacking any fetuslike features, slated for destruction at fertility clinics, remains controversial in the United States. Privately funded researchers can undertake stem cell experimentation without being subject to federal regulation; they may, however, be subject to state regulation.[79] In 1996, the United States Congress passed legislation blocking the expenditure of federal funds for any research or experimentation in which human embryos are destroyed, discarded or knowingly subjected to risk of injury or death. Congress has renewed the ban on federal funding of embryonic research each year thereafter, thereby relegating embryo research to the private sector.[80] In August 2001, President George W. Bush allowed federal funding for research but only on existing embryonic stem cells, so no new embryos would have to be destroyed.[81]

From the viewpoint of the Jewish tradition, if the untransplanted embryos can be destroyed, then research on embryonic stem cells is permissible, as it serves a "common good," namely, preventing or combating disease. Furthermore, under Jewish law, eggs fertilized in vitro, but not implanted, have no standing and may be discarded. The restrictions pertaining to abortion apply only to embryos implanted in a woman's womb. Conversely, embryos outside the womb are mere clumps of cells, having no legal or moral status, and may be discarded.[82] Following this line of reasoning, one Reform responsum concluded that embryos fertilized in vitro, existing outside the womb, are not fetuses; therefore, these excess embryos may be discarded. If they can be discarded, then these excess embryos can be used for research purposes provided they are "handled with the proper respect and reverence."[83]

Furthermore, unused, fertilized eggs may be destroyed or used for research purposes, traditionalist Rabbi Bleich reasons because Jewish law concerns itself only with phenomena that may be seen.[84] Subvisual phenomena are of no consequence. Thus, legal cognizance of an organism only can be taken when it becomes visible to the naked eye. Because an embryo, during its very early stages of development, is subvisual it has no legal existence; the same is true of excess embryos fertilized in vitro.

Under these rationales, the Jewish tradition permits stem cell research to be undertaken on surplus, untransplanted embryos, frozen four to seven days post-conception, that otherwise would be discarded. This permissive position is subject, of course, to the informed consent of the involved individuals or couples to the use of the embryos.

In analyzing the difficult topic of abortion, this chapter has sought to reconcile the competing interests of a mother and her fetus. While permitting abortions prior to the third trimester, because its soul had not yet entered into the fetus's body, Spiritual Judaism would permit a woman to terminate a fetal existence for the most serious of reasons, not because her pregnancy is unwanted or inconvenient. For Spiritual Judaism, a strong reason must exist to destroy a potential person. Abortion ought not to be undertaken lightly or flippantly. Spiritual Judaism thus places great emphasis on couples avoiding unwanted pregnancies, thereby dealing with the problem in the first place. Limiting pre-marital sex as discussed in Chapter 3, couples must accept their responsibilities before having intercourse and not look to using abortion as a technique for post-conception birth control.

Notes

1. Lisa M. Koonin et al, "Abortion Surveillance -- United States, 1997," CDC Surveillance Summaries, *Morbidity and Mortality Weekly Report* 49:SS-11 (December 8, 2000): 1-44; Stanley K. Henshaw, "Abortion Incidence and Services in the United States, 1995-1996," *Family Planning Perspectives* 30:6 (November/December 1998): 263-270, 287, at 264 (Table 1. Number of Reported Abortions, Abortions Rate and Abortion Ratio United States, 1973-1996).
2. Koonin, "Abortion Surveillance -- United States, 1997", 14 (Figure 1. Number, ratio, and rate of legal abortions performed, by year -- United States, 1970-1997), 19 (Table 2. Number, ratio, and rate of legal abortions and source of reporting, by year -- United States, 1970-1997).
3. *Id.*, 5, 27-28 (Table 6. Reported legal abortions, by weeks of gestation and state of occurrence -- selected states, United States, 1997).
4. Aida Torres and Jacqueline Darroch Forrest, "Why Do Women Have Abortions?," *Family Planning Perspectives* 20:4 (July/August 1988): 169-176.
5. James C. Mohr, *Abortion in America: The Origins and Evolution of National Policy, 1800-1900* (New York: Oxford University Press, 1978), 229-230. See generally Leslie J. Reagan, *When Abortion Was A Crime: Women, Medicine and Law in the United States, 1876-1973* (Berkeley: University of California Press, 1997).
6. Mohr, *Abortion in America*, 250-257.
7. *Roe* v. *Wade*, 410 U.S. 113, fn. 37, 140 (1973).
8. 410 U.S. 113 (1973).
9. *Id.* at 163.
10. *Id.* at 163-164.
11. 505 U.S. 833, 860 (1992).
12. 530 U.S. 914 (2000).
13. Basil F. Herring, *Jewish Ethics and Halakhah for Our Time: Sources and*

Commentary (New York: KTAV, 1984), 28; Fred Rosner, M.D., *Modern Medicine and Jewish Ethics*, 2nd edition (Hoboken, NJ: KTAV, 1991), 137; J. David Bleich, "Abortion in Halakhic Literature," *Tradition* 10:2 (Winter 1968): 72-120, at 88, reprinted in *Jewish Bioethics*, ed. Fred Rosner and J. David Bleich (New York : Hebrew Publishing Co., 1979), 147; J. David Bleich, *Contemporary Halakhic Problems* (New York: KTAV, 1977), 348.

14. *The Code of Maimonides*, Book Eleven, Book of Torts, Murder and Preservation of Life, trans. Hyman Klein (New Haven: Yale University Press, 1954), Chapter 1:9, 196-197. See also Rosner, *Modern Medicine and Jewish Ethics*, 140-121; Herring, *Jewish Ethics and Halakhah*, 29, 32.

15. Rabbi Shlomo Ganzfried, *Kitzurn Shulchon Oruch*, vol. II, trans. Rabbi Eliyahu Touger (New York: Moznaim, 1991), Chapter 184:11, 709.

16. David Kraemer, "Jewish Ethics and Abortion," *Tikkun* 8:1 (January-February 1993): 55-58, 77-78, at 56-57.

17. Bleich, *Contemporary Halakhic Problems*, 336-337; Bleich, "Abortion in Halakhic Literature," 79-80, reprinted in *Jewish Bioethices*, 140-141.

18. I have drawn on Bleich, "Abortion in Halakhic Literature," reprinted in *Jewish Bioethics*; Bleich, *Contemporary Halakhic Problems*, 325-371; Fred Rosner, "The Jewish Attitude Toward Abortion," Tradition 10:2 (Winter 1968): 48-71, reprinted in *Contemporary Jewish Ethics*, ed. Menachem Marc Kellner (New York: Sanhedrin, 1978); Herring, *Jewish Ethics and Halakhah*, 25-45; Alex J. Goldman, *Judaism Confronts Contemporary Issues* (New York: Shengold, 1978), 35-62.

19. Bleich, "Abortion in Halakhic Literature," 82-84, reprinted in *Jewish Bioethics*, 143-144, 146; Bleich, *Contemporary Halakhic Problems*, 340-342; Herring, *Jewish Ethics and Halakhah*, 36-38.

20. Bleich, "Abortion in Halakhic Literature," 88-94; reprinted in *Jewish Bioethics*, 148-152; Bleich, *Contemporary Halakhic Problems*, 349-351.

21. Bleich, "Abortion in Halakhic Literature," 94-95, reprinted in *Jewish Bioethics*, 152; Bleich, *Contemporary Halakhic Problems*, 354-355; CCAR Responsum No. 41. "Abortion (1958)," Solomon Freehof, *Recent Reform Responsa* (Cincinnati: Hebrew Union College Press, 1963), 191.

22. Bleich, "Abortion in Halakhic Literature," 96, reprinted in *Jewish Bioethics*, 153; Bleich, *Contemporary Halakhic Problems*, 356.

23. Rachel Biale, "Abortion in Jewish Law," Tikkun 4:4 (July-August 1989): 26-28, at 27.

24. Bleich, "Abortion in Halakhic Literature," 102, reprinted in *Jewish Bioethics*, 158; Bleich, *Contemporary Halakhic Problems*, 363; Herring, *Jewish Ethics and Halakhah*, 42; Immanuel Jakobovits, "Jewish Views on Abortion" in *Jewish Bioethics*, 124; Isaac Klein, "Abortion and Jewish Tradition," *Conservative Judaism* 24:3(Spring 1990): 26-33, at 31-32, reprinted in *Contemporary Jewish Ethics*, 276.

25. Herring, *Jewish Ethics and Halakhah For Our Time*, 42.

26. Bleich, "Abortion in Halakhic Literature," 102, reprinted in *Jewish Bioethics*, 158; Bleich, *Contemporary Halakhic Problems*, 363.
27. Rabbi Kassel Abelson, "Prenatal Testing and Abortion," in *Proceedings of the Committee on Jewish Law and Standards of the Conservative Movement 1980-1985* (New York: The Rabbinical Assembly, 1988), 6-7.
28. Rabbi Robert Gordis, "Abortion: Major Wrong or Basic Right?", in *Proceedings of the Committee on Jewish Law and Standards 1980-1985*, 27.
29. Rabbi Elliot N. Dorff, "*This Is My Beloved, This Is My Friend:" A Rabbinic Letter on Intimate Relations* (New York: The Rabbinical Assembly, 1996), 26, 34; Elliot N. Dorff, *Matters of Life and Death: A Jewish Approach To Modern Medical Ethics* (Philadelphia: Jewish Publication Society, 1998), 56. Rabbis Ben Zion Bokser and Kassel Abelson, "A Statement on the Permissibility of Abortion," in *Proceedings of the Committee on Jewish Law and Standards 1980-1985*, 37 conclude that abortion is permissible if the "continuation of pregnancy might cause the mother severe physical or psychological harm...." In the "earlier" stages of pregnancy, Conservative authorities would allow a therapeutic abortion if there is any threat to the mother's health, directly or indirectly, physically or psychologically. The facts must be established by reliable medical evidence. In the "later" stages of pregnancy, abortion if permitted only when the birth of the fetus presents a direct threat to the mother's life, physically or psychologically, even if the danger is remote. Isaac Klein, "Abortion and Jewish Tradition," 32, reprinted in *Contemporary Jewish Ethics*, 277.
30. Gordis, "Abortion: Major Wrong or Basic Right?", in *Proceedings of the Committee on Jewish Law and Standards 1980-1985*, 24.
31. David M. Feldman, *Birth Control in Jewish Law: Marital Relations, Contraception, and Abortion as Set Forth in the Classic Texts of Jewish Law* (Northvale, N.J.: Jason Aronson, 1998), 290, 292.
32. Abelson, "Prenatal Testing and Abortion," in *Proceedings of the Committee on Jewish Law and Standards 1980-1985*, 9 and Gordis, "Abortion: Major Wrong or Basic Right?", *id.*, 24-26.
33. Bokser and Ableson, "A Statement on the Permissibility of Abortion" in *Proceedings of the Committee on Jewish Law and Standards of the Conservative Movement 1980-1985*, 37; Abelson, "Prenatal Testing and Abortion," *id.*, 9.
34. Robert Gordis, *Love & Sex: A Modern Jewish Perspective* (New York: Farrar Strauss Giraux, 1978), 147.
35. CCAR, Resolution Adopted by the Fifty Third General Assembly of the Union of American Hebrew Congregations (UAHC), 1975, available from UAHC website, www.uahc.org; Resolutions Adopted by the CCAR, 1967, 1975 (86th Annual Convention), and 1980 (95th Annual Convention), available from the CCAR website, www.ccarnet.org.
36. CCAR Responsum No. 16. "When Is Abortion Permitted? (1985)" in Walter Jacob, *Contemporary American Reform Responsa* (New York: CCAR, 1987), 26-27. See also CCAR Responsum No. 155. "The Abortion of an Anencephalic

Fetus (1988)" in Walter Jacob, *Questions and Reform Jewish Answers: New American Reform Responsa* (New York: CCAR, 1992), 253. Responsum, "The Tay-Sachs Program," in Solomon B. Freehof, *Today's Reform Responsa* (Cincinnati: Hebrew Union College Press, 1990), 48-49, would permit the abortion of a fetus affected with Tay-Sachs disease because of parental anguish.
37. CCAR Responsum No. 171. "Abortion (1958)," *American Reform Responsa, Collected Responsa of the CCAR 1889-1983*, ed. Walter Jacob (New York, CCAR, 1983), 543.
38. Balfour Brickner, "Judaism and Abortion," in *Contemporary Jewish Ethics*, 282-283.
39. Rabbi David M. Feldman, "Abortion: The Jewish View" in *Proceedings of the Committee on Jewish Law and Standards 1980-1985*, 17; David M. Feldman, *Birth Control in Jewish Law*, 349.
40. Dorff, "*This Is My Beloved*," 27-28; Dorff, *Matters of Life and Death*, 132-133.
41. CCAR Responsum No. 16. "When Is Abortion Permitted? (1985)," in *Contemporary American Reform Responsa*, 27.
42. CCAR Responsum 5755.13, "Abortion to Save Siblings from Suffering," in *Teshuvot for the Nineties: Reform Judaism's Answers to Today's Dilemmas*, ed. W. Gunther Plaut and Mark Wahofsky (New York: CCAR, 1997), 176.
43. J. David Bleich, *Judaism and Healing: Halakhic Perspectives* (New York: KTAV, 1981), 98-99; Bleich, "Abortion in Halakhic Literature," 104-105, reprinted in *Jewish Bioethics*, 160-161; Bleich, *Contemporary Halakhic Problems*, 366-367; Fred Rosner and Moshe D. Tendler, *Practical Medical Halachah* (Northvale, NJ: Jason Aronson, 1997), 33; Jakobovits, "Jewish Views on Abortion" in *Jewish Bioethics*, 123-124.
44. Bleich, "Abortion in Halakhic Literature," 87, 105, reprinted in *Jewish Bioethics*, 161; Bleich, *Contemporary Halakhic Problems*, 346, 367; Herring, *Jewish Ethics and Halakhah*, 42-43.
45. Bokser and Abelson, "A Statement on the Permissibility of Abortion," in *Proceedings of the Committee on Jewish Law and Standards 1980-1985*, 37; Dorff, *Matters of Life and Death*, 130-131. See also CCAR Responsum No. 16, "When Is Abortion Permitted? (1985)," in *Contemporary American Reform Responsa*, 26-27. Kassel, "Prenatal Testing and Abortion," in *Proceedings of the Committee on Jewish Law and Standards 1980-1985*, 9, added two additional requirements, namely, that the child will be born with major defects precluding a normal life and these defects "make the mother and the family anxious about the future."
46. Dorff, *Matters of Life and Death*, 131-132. However, Feldman in *Birth Control in Jewish Law*, 292, would rule out abortion because it is uncertain whether the fetus would be born defective or how bad its life would be.
47. CCAR Responsum No. 16, "When Is Abortion Permitted? (1985)," in *Contemporary American Reform Responsa*, 27. See also Feldman, *Birth Control*

in Jewish Law, 292, permitting an abortion for a potentially deformed that would child cause his or her mother severe anguish.
48. Dorff, *Matters of Life and Death*, 131-132, 154.
49. I have drawn on Adrienne Asch and Gail Geller, "Feminism, Bioethics, and Genetics," in *Feminism & Bioethics: Beyond Reproduction*, ed. Susan M. Wolf (New York: Oxford University Press, 1996), 335-340; Jeffery Botkin, "Fetal Privacy and Confidentiality," Hastings Center Report 25:5 (September-October 1995): 32-39; Philip Kitcher, *The Lives to Come: The Genetic Revolution and Human Possibilities* (New York: Simon & Schuster, 1996), 285-307.
50. Immanuel Jakobovits, *Jewish Medical Ethics* (New York: Bloch, 1959), 182-183; Feldman, *Birth Control in Jewish Law*, 273-275.
51. Lewis D. Solomon, *The Jewish Tradition and Choices at the End of Life: A New Judaic Approach to Illness and Death* (Lanham, MD: University Press of America, 2001), 68-69.
52. See generally, Martin Berry, "The Development of the Human Nervous System," in *Brain and Behavioural Development: Interdisciplinary Perspectives on Structure and Functions*, ed. John W. T. Dickerson and Harry McGurk (Glasgow: Surrey University Press, 1982), 6-47; N. Herschkowitz, "Brain Development in the Fetus, Neonate and Infant," Biology of the Neonate 54:1 (July 1988): 1-19; J. Korein, Ontogenesis of the Fetal Nervous System: The Onset of Brain Life, Transplantation Proceedings 22:3 (June 1990): 982-983. See also, Michael Tooley, *Abortion and Infanticide* (New York: Oxford University Press, 1983).
53. I have drawn on D. Gareth Jones, "Brain Birth and Personal Identity," Journal of Medical Ethics 15:4 (December 1989): 173-178; D. Gareth Jones, "The Problematic Symmetry Between Brain Birth and Brain Death," Journal of Medical Ethics 24:4 (August 1998): 237-242. See also Raymond D. Adams, Maurice Victor, Allan H. Ropper, *Principles of Neurology*, Sixth Edition (New York: McGraw-Hill, 1997), 574-575; Mark E. Molliver, Ivica Kostović, Hendrik Van der Loos, "The Development of Synapses in Cerebral Cortex of the Human Fetus," Brain Research 50:2 (February 28, 1973): 403-407.
54. Susan Tawia, "When Is the Capacity for Sentience Acquired During Human Fetal Development," Journal of Maternal-Fetal Medicine 1:2 (April 1992): 153-162; J.A. Burgess and S.A. Tawia, "When Did You First Begin to Feel It? -- Locating the Beginnings of Human Consciousness," Bioethics 10:1 (January 1996) 1:26. See generally, John R. Hughes, *EEG in Clinical Practice* (Boston: Butterworths, 1982), 69-94.
55. CCAR Responsum No. 16. "When Is Abortion Permitted? (1985)," in *Contemporary American Reform Responsa*, 27.
56. Stanley K. Henshaw, "Unintended Pregnancy in the United States," *Family Planning Perspectives* 30:1 (January/February 1998): 24-29, 46, at 26.
57. *Id.*, 27; Julia Gorin, "Reinterpreting a Woman's 'Choice,'" *Christian Science Monitor*, August 14, 2000, 9.

58. See generally the classic work on this topic, Feldman, *Birth Control in Jewish Law*.
59. Rosner, *Modern Medicine and Jewish Ethics*, 72-81; Fred Rosner, "Contraception in Jewish Law," Tradition 2:12 (Fall 1971): 90-103, reprinted in *Jewish Bioethics*; Bleich, *Judaism and Healing*, 55-64. See also Rabbi Michael Gold, *Does God Belong In The Bedroom*? (Philadelphia: Jewish Publication Society, 1992), 101-112; Gordis, *Love & Sex*, 125-136.
60. CCAR Responsum No. 156. "Birth Control (1927)," in *American Reform Responsa*, 485-499. See generally, Goldman, *Judaism Confronts Contemporary Issues*, 109-137.
61. Dorff, "*This Is My Beloved*," 26-27; Dorff, *Matters of Life and Death*, 123.
62. Elliot N. Dorff, "Artificial Insemination, Egg Donation and Adoption," Conservative Judaism 49:1 (Fall 1996): 3-60, at 47; Dorff, *Matters of Life and Death*, 129-130; Aaron L. Mackler, "An Expanded Partnership With God?: In Vitro Fertilization in Jewish Ethics," *Journal of Religious Ethics* 25:2 (Fall 1997): 277-304, at 289.
63. Joseph G. Schenker, Shaul Yarkoni, Menachem Granat, "Multiple Pregnancies Following Induction of Ovulation," *Fertility and Sterility* 35:2 (February 1981): 105-125, at 106-111.
64. Joyce A. Martin, Marian F. MacDorman, T.J. Matthews, "Triplet Birth: Trends and Outcomes 1971-1994," in *Vital Health Statistics*, Series 21, Number 55 (Hyattsville, MD: National Center for Health Statistics, 1997), 7, 10; B.J. Botting, I. MacDonald Davies, and A.J. MacFarlane, "Recent Trends in the Incidence of Multiple Births and Associated Mortality," *Archives of Disease in Childhood* 62:9 (September 1987): 941-950.
65. Schenker, Yarkoni, Granat, "Multiple Pregnancies," 114-117.
66. Rosner and Tendler, *Practical Medical Halachah*, 35-36; Rosner, *Modern Medicine and Jewish Ethics*, 159. See also Fred Rosner, "Pregnancy Reduction in Jewish Law," The Journal of Clinical Ethics 1:3 (Fall 1990): 181-186, at 185.
67. Richard V. Grazi and Joel B. Wolowelsky, "Multifetal Pregnancy Reduction and Disposal of Untransplanted Embryos in Contemporary Jewish Law and Ethics," *American Journal of Obstetrics and Gynecology*, 165:5(Part 1) (November 1991): 1268-1271, at 1270, reprinted in *Be Fruitful and Multiply: Fertility Therapy and the Jewish Tradition*, ed. Richard V. Grazi (Jerusalem: Genesis Jerusalem Press, 1994), 178.
68. J. David Bleich, "Pregnancy Reduction," *Tradition* 29:3 (Spring 1995): 55-63, at 56; J. David Bleich, *Bioethical Dilemmas: A Jewish Perspective* (Hoboken, NJ: KTAV, 1998), 271.
69. Rosner, *Modern Medicine and Jewish Ethics*, 159; Rosner, "Pregnancy Reduction," 185.
70. Aaron L. Mackler, "An Expanded Partnership With God?", 288. This responsum was approved by the Conservative Movement's Committee on Law and Standards. Dorff, *Matters of Life and Death*, 55.

71. Richard V. Grazi, Joel B. Wolowelsky, Raphael Jewelewicz, "Assisted Reproduction in Contemporary Jewish Law and Ethics," Gynecologic and Obstetric Investigation 37:4 (May 1994): 217-225, at 220-221; Grazi and Wolowelsky, "Multifetal Pregnancy Reduction," at 1270, reprinted in *Be Fruitful and Multiply*, 178-181; Rosner, "Pregnancy Reduction," 184-185.
72. Bleich, "Pregnancy Reduction," 59; Bleich, *Bioethical Dilemmas*, 275.
73. Bleich, "Pregnancy Reduction," 60-62; Bleich, *Bioethical Dilemmas*, 276-278; Yitzchak Mehlman, "Multi-Fetal Pregnancy Reduction," Journal of Halacha and Contemporary Society 27 (Spring 1994): 35-68, at 55-59.
74. Mackler, "An Expanded Partnership With God?", 288.
75. Grazi and Wolowelsky, "Multifetal Pregnancy Reduction," at 1270, reprinted in *Be Fruitful and Multiply*, 181.
76. Dorff, *Matters of Life and Death*, 129-130.
77. *Id.*; Mehlman, "Multi-Fetal Pregnancy Reduction," 63-64.
78. National Institutes of Health, *Stem Cells: Scientific Progress and Future Research Directions* (June 2001); Nicholas Wade, "Teaching the Body To Heal Itself," *New York Times*, November 7, 2000, D1.
79. The array of state rules are summarized in Lori B. Andrews, "State Regulation of Embryo Stem Cell Research," in National Bioethics Advisory Commission, *Ethical Issues in Stem Cell Research*, volume II (Rockville, MD: National Bioethics Advisory Commission 2000). See also Sheryl Gay Stolberg, "Washington Not Alone in Cell Debate," *New York Times*, July 23, 2001, A12.
80. Ellen J. Flannery and Gail H. Javitt, "Analysis of Federal Laws Pertaining To Funding of Human Pluripotent Stem Cell Research, in *Ethical Issues in Stem Cell Research*, volume II, D-6 to D-9.
81. "Bush's Address on Federal Financing for Research With Embryonic Stem Cells," *New York Times*, August 10, 2001, A16.
82. Mehlman, "Multi-Fetal Pregnancy Reduction," 42-43; Grazi and Woloweksky, "Multifetal Pregnancy Reduction," 1269, reprinted in *Be Fruitful and Multiply*, 180.
83. CCAR Responsum No. 5757.2. "In Vitro Fertilization and the Status of the Embryo," available from the CCAR website, www.ccarnet.org.
84. Bleich, *Bioethical Dilemmas*, 211.

Chapter 10

Conclusion

We need the courage and wisdom to take and reshape the ancient ideals of the Jewish tradition to deal with the challenges and modern realities of sex and procreation. As we come to the end of this book, three themes are noteworthy: the changing social and familial role of women; the emerging diversity of families; and the heightened role of technology in human reproduction.

First, we are in the midst of a fundamental reexamination of our assumptions and notions of appropriate gender roles, specifically, the role and status of women not only in the family but also, more generally, in society. The Jewish tradition has long assumed that wives raise children and keep house; that they are child bearers and rearers as well as housekeepers; while husbands take on the role of economic providers, bosses at work and at home, and protectors. The past three or four decades have witnessed a transformation of the relationship between men and women in the United States, enabling both genders to participate in the entirety of life, at work and in the family.

Women have entered the work force in record numbers. Women as well as men want the freedom and opportunities found in the workplace. Women have children later in life, enabling them to make a greater and earlier investment in job skills. Those with careers have become less dependent on men for financial support. As women grew more self-sufficient, gaining economic power, with their own careers, incomes, cars, and homes, they came to have greater bargaining power in marriage.

Given the high incidence of divorce, American women have come to realize they cannot base their future expectations on the assumption of being supported by their husbands for the rest of their lives. During the last thirty years, as more women have become economically self-sufficient without husbands, they no longer had to endure unhappy marriages.

Spiritual Judaism urges the need to turn aside from the unacceptable sexist assumptions of the Jewish tradition, the rigid, preconceived gender roles -- shaped by men, for men -- over the millennia. These assumptions, often failing to recognize the injustice of the power imbalance between males and females, are based more on social attitudes about gender differences than on the institution of marriage. They have led to the creation of rules and precedents constructed and applied without female participation resulting in the subordination of women. We are, however, in the process of re-imaging family relationships, based on equality and partnership, not male domination and female submission.

Spiritual Judaism supports the trend from a patriarchal to a more egalitarian culture, marked by a greater degree of equality in marriage, less gender differentiation in roles, and a sharing of family and work life. With the advent of two career couples, came a breakdown in traditional roles of married men and women; they now share more equally domestic chores rather than allocating all to one partner. Men are (and ought to be) involved in raising children in two-parent households. Both husbands and wives have come to assume more equal responsibility for child care and, more generally, for the whole parenting process. Wives can be domestic without a betrayal of themselves; they can have a connection to a husband without submission.

Spiritual Judaism advocates an open path for each woman's life. Opposing any limits on what a woman can aspire to, Spiritual Judaism recognizes, depending on individual psyches, that the levels of female ambitions may change over time. Women have a range of ambition levels. These fluctuate throughout their lives. Those with high ambition levels, who devote themselves to their independence and careers, think about having children in their mid to late 30s. Obviously, frustrations arise when ambition levels do not dip as their desires for children surge. Those with lower ambition levels tend to want children in their 20s.

A certain proportion of women (and men, for that matter) lack the temperament for being a successful parent. Some women should not even have children; others ought not stay home all day with their kids. Thus, it is desirable for women to pursue different paths that may or may not depend on marriage, children, or work. There is no need to assume or expect a sameness in female motivation or outcomes.

In the twenty-first century, more and more of us will view marriage as an emotional and economic partnership, where wives are no longer subordinated to their husbands or seen as resources for male subjugation. Spiritual Judaism recognizes that we can reject the traditional sex roles and gender stereotypes that forced women into constrained roles, without the moral fabric of society crumbling.

Second, a need also exists for the Jewish tradition to reconceptualize its notions of sex and the family. Based on the interpersonal virtues of love and compassion, Spiritual Judaism recommends not limiting sex or reproduction to what tradition regards as "natural," namely, husband-wife intercourse. As we witness the setting up of more non-traditional families, many of us have come to realize that there is no longer one natural process to marriage, family, or conception. We can go beyond the traditional nuclear family, consisting of a husband-breadwinner, a wife-housekeeper, and their genetic children. A child need not have two complementary married biological progenitors.

Welcoming the major transformation of the family and sexuality occurring at this time, Spiritual Judaism supports a diversity of family forms, including single-parent, heterosexual and homosexual couples, the raising children conceived by artificial insemination by donor and/or surrogacy, cohabiting partners, families blended by divorce and remarriage, provided each unit is spiritually-based, life-affirming, and consisting of faithful partners striving to facilitate the attaining of fulfilling lives for each family member.

Marriage and family structures are diverse, and, in the future, will become even more varied. Their form depends on need and circumstance. Judaism should not block their development and evolution. The Jewish tradition needs to take a pluralistic view of marriage, including public proclaimed, legally recognized same-gender unions. For Spiritual Judaism marriage for homosexual couples does not demean marriage for heterosexuals. Furthermore, Spiritual Judaism does not accept the view that the emerging variety of family forms signals a disastrous breakdown in our social order.

Spiritual Judaism also questions the traditional assumptions about sexuality. Permanently committed (and in the future, solemnized) homosexual relationships are not a perversion of the natural order, but a way in which some express their sexual desires and their love. The forms in which human sexual needs are met are diverse, culturally variable, and legitimate in the context of caring, monogamous relationships.

New family configurations, particularly same-sex unions, are pro-family and pro-children. If based on enduring commitments, they provide the economic benefits of the traditional family unit and offer barriers to relational breakups. In sum, the traditional nuclear family is not the only ideal for us to emulate.

Third, over the past three or four decades changes regarding sex, procreation and the family have been driven not only by the transformation of the economy as well as the evolution of moral and cultural standards but also by shifts in technology. We have witnessed artificial insemination, test tube babies, surrogate mothers, and some day, may witness cloning and genetic engineering. The physician and the geneticist have become active partners in the creation of new human life.

If a couple is infertile because a husband cannot provide sperm or the wife is unable to produce a fertilizable egg or carry a fetus to term, they could remain childless, adopt, or turn to some form of assisted reproduction, such as artificial insemination, in vitro fertilization, or surrogacy. Because not everyone wants to merge his or her genes with those of a stranger (whether egg or sperm donor), cloning, when it is perfected, will offer another option. Desperate individuals and couples will encourage researchers and clinicians to push the envelope of what is possible.

Assisted reproductive technologies, surrogacy arrangements, and cloning provide new choices for those within and outside the nuclear family structure. These technologies expand the reproductive options for unmarried individuals and homosexuals. Also separating reproduction from sex, assisted donor insemination, now, and cloning, in future, free women from dependency on men. Modern technologies eliminate a significant barrier to female reproduction, previously created by the need to engage in sexual intercourse with a male.

Spiritual Judaism supports the use of all modern technological means to produce a child, if implemented in a loving, compassionate and responsible manner. It accepts a technological approach to human reproduction, separating procreation from sex.

Modern medicine will also push back the bounds of procreational opportunity for women. New techniques may considerably extend the reproductive period females will enjoy. In the twenty-first century, along with longer life spans for everyone, we will need to face the question of whether having children is a major value for "older" women. Is having children the only thing for 50, 60, or 70 year olds to do with extra years of life, the same things women did when they were in their 20s and 30s?

There are and, in the future, will be many paths to reach the goal of having a child. The key is the love a parent (or parents) provide after birth and the assurance that a child is well carried for, whether or not his or her parents are married or are of different (or the same) genders.

As we come to the end of this book, it is important to understand that Spiritual Judaism rests on the notion that creation is an ongoing process in which humans are called to active participation, as co-creators, with God. Human destiny evolves over time, continually being recreated. Spiritual Judaism urges us to remain open to human creativity using technological innovations, implemented in a loving, compassionate, and responsible manner.

Selected Bibliography

Borowitz, Eugene B. *Choosing A Sex Ethic: A Jewish Inquiry.* New York: Schocken, 1969.

Dorff, Elliot N. *Matters of Life and Death: A Jewish Approach to Modern Medical Ethics.* Philadelphia: Jewish Publication Society, 1998.

_____. "This Is My Beloved, This Is May Friend": A Rabbinic Letter on Intimate Relations. New York: The Rabbinic Assembly, 1996.

Ellenson, David H. "How to Draw Guidance from a Heritage: Jewish Approaches to Mortal Choices," in *Contemporary Jewish Ethics and Morality: A Reader.* Edited by Elliot N. Dorff and Louis E. Newman. New York: Oxford University Press, 1995.

Solomon, Lewis D. *Jewish Spirituality: Revitalizing Judaism for the Twenty-First Century.* Northvale, N.J.: Jason Aronson, 2000.

_____. *The Jewish Book of Living and Dying.* Northvale, N.J.: Jason Aronson, 1999.

_____. *The Jewish Tradition and Choices at the End of Life: A New Judaic Approach to Illness and Dying.* Lanham, Maryland: University Press of America, 2001.

Index

www.ingramcontent.com/pod-product-compliance
Ingram Content Group UK Ltd.
Pitfield, Milton Keynes, MK11 3LW, UK
UKHW012034150926
475560UK00001B/7

9 780761 822271